MW01087837

The Neck

The Neck

A Natural and Cultural History

Kent Dunlap

UNIVERSITY OF CALIFORNIA PRESS

University of California Press
Oakland, California

Cataloging-in-Publication data is on file at the Library of Congress.

ISBN 978-0-520-39303-5 (cloth : alk. paper)
ISBN 978-0-520-39305-9 (ebook)

Manufactured in the United States of America

33 32 31 30 29 28 27 26 25 24
10 9 8 7 6 5 4 3 2 1

For Terri, who loves morphology

CONTENTS

ACKNOWLEDGMENTS

Many people provided all sorts of help at different phases of writing this book. I benefited greatly from early conversations with James Trostle, Jessica Chotiner, Daniel Blackburn, Morgan Lloyd, Beth Casserly, Peter Kyle, and Devon Treadway. Mary Mahoney and my agent, Michelle Tessler, helped me formulate the book proposal. Several people offered their expertise and perspectives through interviews: Rachnya Ramya, George Pearlman, Joanne Scattergood, Scott Raymond, Lee Stang, and Diana Hews. These folks contributed their expertise through conversations and reviewing sections of the manuscript: Douglas Johnson, Stephen Rockel, Carl Malchoff, Jeff Podos, Chris Sidor, Mathew Wedel, Shane Ewegen, Elli Findly, Tobias Riede, Jessica Feeley, Jonathan Elukin, Gabe Hornung, Leo Fleishman, Donald Dearborn, Kari Theurer, Zoe Maggioni, and Brett DiBenedictis. Others generously provided editorial comments on large portions of the book: Mark Davis, David Schonfeld, Gretchen Hathaway, Sean Coco, Alexandra Soiseth, Michelle Tessler, and my editors at the University of California Press, Chloe Layman, Stacy Eisenstark, Chad Attenborough, and Amy Smith Bell. I am very grateful to these interlocutors, experts, and editors.

I had the great fortune of growing up in a family full of curiosity, and it is easy to trace my broad interests in nature and culture to these cherished people: my parents, Jean and Harold, and my siblings, Susan and Paul Dunlap. My own children, Ruth, Luke, and Samuel, have always asked a lot of great questions, as kids and adults, and their curiosity inspires me to continue asking and learning.

My greatest gratitude goes to Terri Williams, my favorite conversationalist about necks—and everything else—my best writing critic, my dictionary on two feet, my persistent visionary, my wife. Thank you again and again.

PREFACE

Vital & Vulnerable

All of us live lives full of vitality and vulnerability. But by many measures, Isadora Duncan (1877–1927) led a life more extreme. As a choreographer, Duncan revolutionized the world of dance. Even beyond dance, she was an iconoclast in most every way. She broke almost every convention of society, until her flamboyant flowing red scarf abruptly broke her own neck.

Duncan's unconventional life began early. She quit school at age ten, finding it too restrictive. Within a few years she was earning money teaching dance in California. In her twenties she traveled around Europe and created a new form of dance that revolted against the rigidity of ballet and emphasized free-flowing and expressive movements of nature. In the ballet tradition that preceded Duncan, dancers elevated their heads as erect extensions or gentle arcs above the spine, but in Duncan's new dance they commonly extended their heads backward, flexed them forward, and twisted at their necks in an expressive spectacle. In Duncan's piece "Revolutionary," the solo dancer kneels on stage and alternates between thrusting her head downward in rage and despair and arching it upward, desperately pleading for help or explanation from above. Duncan's personal life was no less revolutionary. She cavorted with bohemians across European capitals. She bore three children outside of marriage. She was at various times an outspoken feminist, bisexual, atheist, and communist. In dance and in life Duncan was powerfully expressive at the neck, both in movement and voice.

At 9:40 p.m. on September 14, 1927, Duncan draped herself in a crimson scarf "twice her size" and hopped into a convertible in Nice, France. "Goodbye, my friends, I'm off to glory!" she reportedly exclaimed. As the car pulled away, the long scarf tangled in a rear wheel, tightened, pulled her from the car, and fatally snapped her neck. Duncan's big life was gone instantly, finished off at the expressive but vulnerable constriction of her neck.[1]

• • •

In one small region—less than 1 percent of the body—the neck concentrates both the vitality and the vulnerability of the human condition. All our head movements, which so potently express our attitude and focus of attention, are controlled by the neck's shortening muscles. All our utterances, both meaningful and mundane, originate in its quivering vocal cords. All our body movements and sensations depend on electrical signals passing through its spinal cord and nerves. The brain is fed blood through its pulsing vessels, and the body is fed air and food through its windpipe and esophagus. Our necks work all the time to express and sustain ourselves.

Well, almost all the time. In rare and unpredictable moments the neck fails. This fallibility is the disturbing, sometimes terrifying condition of human life, and it is the dark counterpoint to the phenomenal capabilities of the neck. The thinness that makes the neck so flexible also makes it easy to break. The tubes that are so vital to its function are narrow and near the surface, making them easy to clog or puncture. In the rare occasions when something doesn't work, we could be moments from death: paralyzed, suffocated, or exsanguinated. Consciously or unconsciously, we deeply sense this precarious vulnerability at the neck, and, to some degree, we carry it in the back of our minds. This book explores the uneasy fusion of the expressive and functional capacity of the neck—its vitality—on one side with its extreme vulnerability on the other. Indeed, I argue that in its anatomy, as in much of life, vitality and vulnerability coexist inseparably side by side.

• • •

On the shelf above my desk sits a book that my dad gave me after his retirement. The cover illustration showing a side view of a human head is both expressive and declarative. Above the jaw you see the face tilted back and away in an enigmatic pose: restful, submissive, and haughty, all at once. The facial features are fine and sparsely sketched. The illustration is captivating because the unusual posture and minimal lines simply invite questions. Below the jaw the illustration is quite different. The model's skin is completely removed to reveal the underlying anatomy of the neck. The impressively detailed illustration shows every strap, tube, cord, and lump that is woven and packed into the side of the neck. Here, the illustration is declarative and explicit. It looks almost like a blueprint showing the complex wires, pipes, and shafts of a building. The drawing is mechanical and also ornate. It is densely informative but still nowhere easy to understand.

The illustrator was Henry Vandyke Carter, and the book is one of the most published medical references in history: *Gray's Anatomy.*[2] I would have loved to have been in the room when the publishers were deciding about the cover illustration. Which of Carter's 363 exquisite engravings would best pull in readers and render the essence, or at least a singularly compelling portion, of the human body? To me, the illustration encapsulates the nature of human bodies and indeed of all organic forms: their aesthetic potency and their material complexity are all wrapped together. Beautiful, evocative, expressive, and also mechanical, physical, and infinitely detailed.

My father was a surgeon (hence the gift of his *Gray's Anatomy*) and my mother was a lover of animals. My long interest in anatomy and my career as an animal physiologist no doubt originated in this biology-centered home. Both my parents, at different times and for different

Figure 1. (a) Dancer and choreographer Isadora Duncan, ca. 1906–1912. Photograph by Charles Ritzmann (Wikimedia).

reasons, had neck surgery and their ailments gave me a particular awareness of the neck's fragility. But my interest in the neck came largely from two entirely different sources. I am a potter by avocation and an improviser by temperament. While I was teaching comparative anatomy labs in graduate school, I was also making pots in my backyard. I made all sorts of vessels (bowls, plates, cups), but at one point I

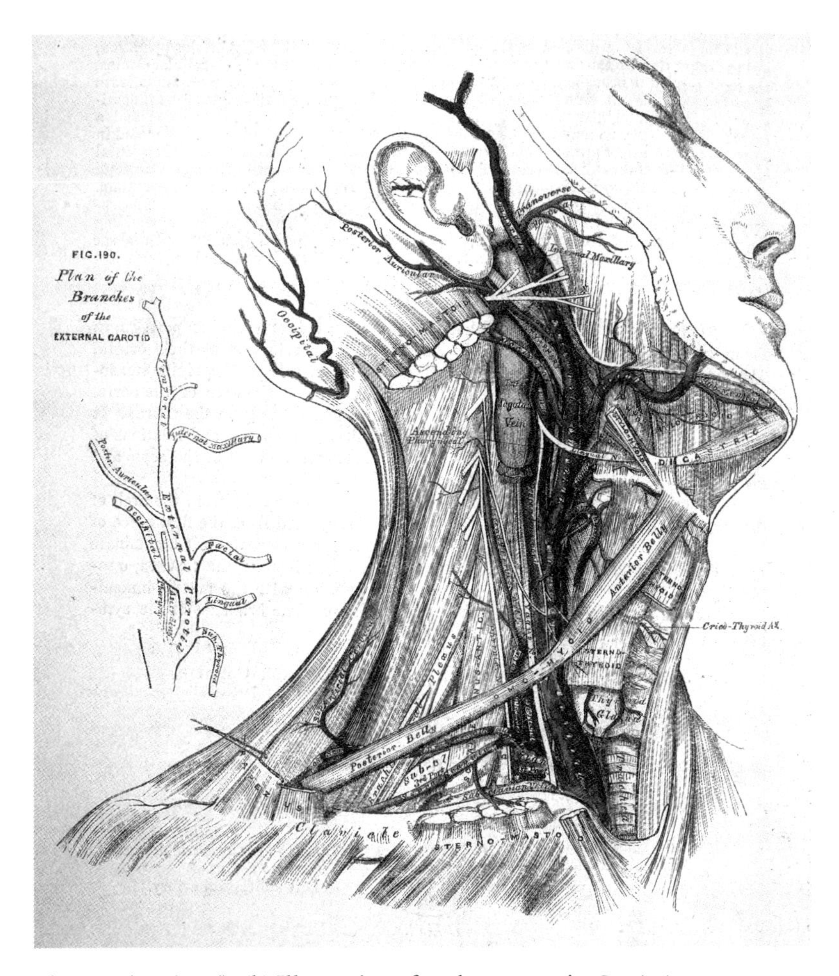

Figure 1 *(Continued)*. (b) Illustration of neck anatomy in *Gray's Anatomy*. Illustrated by Henry Vandyke Carter, ca. 1858 (Gray 1988).

became captivated by vases—that is, pots with necks. It's not just me. Across ages and cultures, vases have been especially pleasing to human sensibilities. They are by far the most common pottery form displayed in art and anthropology museums, in palaces, in houses of worship, and in living rooms. The constriction that defines a vase surely has function for facilitating poured liquids, sealing off the contents, or stabilizing bouquets. But, in parallel, this constriction imbues pots with a curvature and proportion that many call graceful. Bowls, plates, and cups have many pleasing shapes, but they are rarely called graceful.

Like the necks of vases, the necks of humans (and many other animals) are also a locus of aesthetic attention and expression. Many people admire the graceful necks of Audrey Hepburn and elegant trumpeter swans. People decorate their necks in endless ways (jewelry, scarves, ties, collars) or spritz them with perfume. We continuously shift our necks while snapping selfies to get that just-right pose. What is it about this constriction, in bodies and in pots, that so calls us to give the neck attention, to admire it and adorn it? The necks of vases and bodies share a more tragic side as well: they are both sites of extreme vulnerability. Ding a vase on its long neck or torque a human neck too far and, abruptly and irreversibly, it's all over. Vulnerability is the backdrop of their grace. This paradox—part disturbing, part beautiful—set me off to look across cultures and disciplines to see how people at different times and places have reckoned with this unavoidable feature of the human condition.

I was also drawn to the neck because, although its form is refined and graceful, its construction is gloriously quirky and seemingly improvised. I am not a very good planner. My carpentry projects often fall a half-inch short. I am drawn to clay because it is malleable and forgiving, at least when it is wet. You start with a vague idea and then see what happens. On the potter's wheel you can continually revise an object's shape. In ceramic sculpture you can smudge on a new glob of clay, trim off another glob, flatten it, squeeze it, and cut it. In the end the sculpture is an amalgam of additions, deletions, and modifications; some of these choices are apparent in the final product, and others remain hidden.

The human neck is similarly an amalgam. On the outside it looks mostly like a simple, elegant column. But when the skin is drawn back, like on the cover of *Gray's Anatomy*, the neck is revealed as an odd conglomeration of parts all jammed together in a tight space. It's as if the creator of the neck were a ceramic sculptor, not a carpenter. When you learn about how the neck is constructed during embryonic development and evolution, this hunch is confirmed. In both the early embryo and primitive vertebrates (backboned animals), blocks and strips of tissues are laid out in orderly, repeated parts. Then, as embryonic development or evolution moves forward, the invisible hand of the sculptor starts to work. Some of these parts combine, while others move around, change shape drastically, or disappear altogether. New parts pinch off from old parts and other new parts migrate in. It's a jumble in motion, an improviser's delight. Somehow, in the end, the neck that is formed is structurally ornate, functionally versatile, and, at least to me, endlessly curious.

This book is my attempt to make sense of the neck in all its potency, fragility, and quirkiness. Humans are fated to live in a body in which one small region has such an oversized role in our biology, psyche, and culture. What exactly does the neck do for us, and how do we respond, both individually and culturally, to the possibilities and tribulations that it engenders? Just as our highly mobile necks allow us to view and appreciate the world from a range of directions, we can gain a richer, more complete perspective on the neck through multiple approaches, through the sciences as well as the arts and humanities. So this tour of the neck will take us through the anatomy lab, the art and anthropology museums, the clinic, the dance studio, and the music hall. I aim to pull back the skin and disclose the incredible complexity manifested in this one small chunk of our bodies and its ornate workings that sustain daily life. At the same time, I elaborate the multitude of ways humans move, speak, sing, and decorate the neck to express their inner lives and sensibilities. The detailed mechanics and the expressive gestures, all wrapped together.

. . .

Necks emerged in our early terrestrial ancestors about 375 million years ago, and over time they have proven to be remarkably versatile in our animal ancestors and cousins. For everything we can do with our necks, there are animals out there that surpass us in most every way. Owls can twist their heads in a full circle. Bighorn sheep can clash heads with a force that would crush our vertebrae. Snakes can swallow bites that are many times the diameter of their necks. Trumpeter swans can project their voice over a mile (1.6 kilometers). Animals can do things with their necks that are not even in the human repertoire. Turtles can pull their head completely inside their body. Pigeons can make a "milk" in their throat to feed their chicks. Many animals can communicate by flashing or inflating colorful patches of their throats.

The extravagant diversity of necks among animals testifies to the creative capacity of evolution. All vertebrates start embryonically with the same basic layout of clay chunks, and the sculptor goes crazy elongating, squishing, combining, and removing. Occasionally a new variation of the sculpture works a bit better for a particular lifestyle, and it passes on to the next generation. Run this for hundreds of millions of years, and the result is a splendid array of neck forms and a correspondingly fantastic collection of neck functions. To explore this diversity, this tour of the neck meanders through the natural history museum, the zoo, and the many wild places where animals spend their days.

The book begins by confronting the fundamental questions: "Why the neck? What merits all the extreme vulnerability at the neck?" I unpack the anatomy of the neck to reveal its many vital physiological functions: holding and moving the head; transporting air, food, and blood to feed every cell in the body; and secreting hormones that regulate every cell in the body. While the neck connects the head and torso, it also serves to connect creatures to one another through its crucial role in social communication. I discuss how humans and other animals use structures and behaviors at their necks to outwardly express their interior worlds. This includes vocalizations broadcast from the throat (words and songs) as well as flamboyant visual ornaments displayed around the neck (jewelry, collars, and tattoos in humans; manes, frills, and color patches in animals). I delve into the dark world of how humans and animals direct violence at the neck to dominate their subjects or their prey, and, on the brighter side, how they seek protection from threats to the neck through human rituals (e.g., wearing amulets) and unique animal behaviors (e.g., retracting the head into their shell). I highlight examples of how the neck is the site of heroic daily operations and exuberant expression as well as a site of frailty and suffering. By the end, you might never look at a ballerina, a wrestler, a swan, or a lion—or even in the mirror—in the same way.

Humans cannot help but feel uneasy about the concentrated vulnerability at the neck. But let's remember to give it accolades. "Accolades," in contemporary times, refers to honor and praise for meritorious service. The term refers to the bygone ceremony of conferring knighthood by tapping the honored one with a sword on either side of the neck. Originally this word had another connotation. Derived from the Latin *acollare*—the same root that gave us "collar"—it meant to embrace and kiss the neck. So, with affection as well as praise, let us embrace both the vitality and the vulnerability of this small but potent segment of the body.

CHAPTER ONE

Why & What

The Rationale and Contents of the Neck

Imagine these few seconds of your morning ritual: Sitting upright, you tilt back your head to gulp down the last of your coffee or tea. You take a breath, and lower your head back down to your newspaper, book, or smartphone. Your spouse walks in the room. Lifting your head, you locate her or him for a glance and launch some vibrations through the air: "Good morning." You might repeat this sequence most mornings of your adult life, and a version of it is enacted perhaps billions of times daily across the world.

As mundane and automatic as these actions may seem, they rely on the precisely timed operation of dozens of structures in the neck. Muscles shorten, joints twist, flaps open and shut, air and liquids flow, cords and cavities vibrate. In almost every second of our waking hours, we use our neck in so many ways, and usually several of these functions happen at the same time. In almost all cases we generate this coordinated commotion without a conscious thought. Now, imagine your gulp goes down the wrong way, and you lurch forward, coughing and gasping. It is rare but terrifying. Although the neck is a masterful multitasker, it has many major drawbacks. It makes us vulnerable to a quick death from choking or bleeding or permanent paralysis from spinal cord injury. What could possibly be worth all this vulnerability? What's the good in it? In short, why the neck?

THE RATIONALE

Biologists almost always have an initial answer for the "why" of a body part: because the ancestors had it. Humans have a neck because our distant ancestors "invented" it, and we inherited it. Give praise—and blame—to your heritage. The fishlike ancestors of vertebrates had no necks, and neither do their living fish descendants. In early aquatic vertebrates the skull was fused directly to the forelimb (pectoral) skeleton. When vertebrates evolved onto land at the fish-to-amphibian transition, this bony connection was lost. The head became detached from the limbs, and it could move independently. This new head mobility was evident in a key "missing link" fossil, *Tiktaalik*, which dates the origin of the neck to about 375 million years ago.[1]

Our understanding of the evolutionary advantage of the neck during this transition was synthesized by Carl Gans, a prominent comparative anatomist in the second half of the twentieth century.[2] For Gans the fundamental advantage of the vertebrate neck is that it partially separates the body's systems of sensing and locomotion. Put simply, it allows animals to look one way while moving in another direction or to look all around themselves without moving at all. Gans begins his argument by positing that the basic vertebrate body plan, originating about 500 million years ago, evolved primarily through selection to become a highly mobile predator. These earliest vertebrate ancestors—fishlike predators—survived because, as part of their predatory design, they positioned many of their senses in the head and powered fast, long-distance foraging movements with swimming muscles in the tail. With senses and mouth at the front end, they could both detect and grab prey when they first encountered them.

These earliest vertebrates and their fish descendants have no neck, and the sensory head and the locomotory body are combined as one unit. This necklessness has been thoroughly adequate for them; fish are by far the most abundant and species-rich class of vertebrates. In fact, a neck could be detrimental because the fish would have to expend muscular effort to keep its head and body in a single streamlined, hydrodynamic unit in the water. However, necklessness is also a great limitation. Without a neck an animal cannot widely scan the environment or direct its head toward a prey without moving its entire body.[3] For a fish to see a prey swimming below, it must swim to reorient its whole body downward. Similarly, capturing a prey often requires lining up the whole body for the predatory strike. Such movements not only take energy, but they also make the fish more detectable to the target prey.

When vertebrates evolved onto land, the air surrounding them imposed fewer hydrodynamic constraints, and their heads could move separately from their trunks without drastically compromising locomotion. So, according to Gans, the fundamental rationale for the origin of the neck is that it allowed animals to turn their heads to scan a wide panorama and to capture food that was not directly in front of them without moving their whole body. Thus eagles can fly forward while bending their heads downward to search the ground for their next meal. Lizards can detect insects walking to their side, turn their head, and snap up their prey while barely moving a limb. Tortoises can graze a great arc around themselves without moving and lifting their heavy shells.

A second feature of terrestrial life that influenced the evolution of the neck was the possibility of long-distance vision. Light travels much better in air than in water, so when vertebrates colonized land, they had the potential to see objects across much farther distances. However, without necks terrestrial vertebrates would be relegated to short-distance vision because their eyes would be so close to the ground (unless they could fly or climb). Indeed, most amphibians—with their rudimentary, single-joint neck—are restricted to what they can see at ground level. But many reptiles, birds, and mammals have necks long enough to raise their heads above the ground and turn them around, enabling them to peer farther across the landscape. For instance, picture a zebra grazing with its head down, then elevating its head above the tall grass to scan the horizon for predators. Biologists Malcolm

Figure 2. Illustrations of (a) lizard, (b) goose

Figure 2 *(Continued).* (c) horse showing necks that differ in size, proportion, and orientation. Illustrations published by Pearson Scott Foresman.

MacIver and Barbara Finlay have estimated that the 360-degree visual range made possible by a mobile neck and eyes combined with the greater transmission of light in air than in water meant that a terrestrial vertebrate could see objects in a volume of space that is about a millionfold greater than that of an aquatic vertebrate.[4]

After their stubby origins in amphibians, necks diversified widely during evolution, becoming highly mobile and elongated among reptiles, birds, and mammals. Given this diversity of neck forms, the question becomes not simply "why have a neck?" but rather, "why have a neck *of a particular shape?*" Picture the neck shapes of three familiar animals: a lizard, a goose, and a horse. The necks of all these animals enable them to forage without moving their bodies and to see a wide swath around them, yet their necks vary greatly from each other in proportion and function. Many of the differences among their necks (and the necks of all vertebrates) are driven largely by two broader issues that tie the neck to evolutionary modifications outside the neck: how far the animal holds its head above the ground (low or high), and how the animal processes food (swallow it whole or chew it).

Lizards, like most other reptiles, hold their head low and near the ground. They grab prey with their jaws and swallow their prey whole, so they don't need big teeth, jaws, or muscles for chewing. Their necks are mobile but relatively short; they don't need a long neck to reach their head to the ground and capture prey. Because they do not have a heavy chewing apparatus, the heads of lizards are relatively light, and their neck muscles are correspondingly slight. Geese and other birds hold their heads high. They stand bipedally on their hind limbs and devote their front limbs to flying. This upright posture requires long necks to grab objects from the ground with their jaws. To further accommodate flight, they evolved heads and jaws even lighter than those of reptiles, abandoning teeth and heavy chewing muscles altogether. By contrast, horses and other large mammals generally chew their food with massive teeth, jaws, and muscles, and they stand on four legs elevating their heads high above the ground. So their necks must be long enough to reach the ground and strong enough to move their heads up and down. These three examples illustrate that, although the evolutionary origins of the neck may have been to partially separate the senses and feeding anatomy from locomotion, its subsequent diversification in shape and strength was closely integrated into the evolution of posture and feeding mode.

Like geese, we humans are bipedal, yet our necks are not long; like horses, we chew our food, yet our necks are not massive. Human necks were influenced strongly by yet another adaptation outside the neck: our manual dexterity. Along with other primates we are liberated from the necessity of a long neck to reach the ground because we generally grasp objects with our hands rather than our jaws. As fully bipedal primates, humans largely abandoned the use of arms and hands for locomotion. With our upright posture we balance our heads atop our spine without needing strong muscles to hold and move our heads. As a consequence of our bipedality and manual dexterity, humans have really unusual necks compared to most of our vertebrate ancestors and relatives: ours are short, thin, and vertical.

So how can we answer the question why do humans have necks? Our complex history as animals requires a series of answers. As products of evolution, we have necks because our early terrestrial vertebrate ancestors had them. They used necks for moving their heads around to grab objects with their mouths, for elevating their heads, and for expanding their sensory range. Why do we have *our particular* human neck? Our hands have evolved to grab objects and our vertical bodies now elevate the head, but we still use our necks to expand our sensory range, by twisting around. Humans also evolved a unique voice box to serve in our elaborate system of vocal communication. At the neck we turn and talk, twist and shout.

• • •

In contemporary times we use the ideas of evolution and the language of science to answer the question why the neck? But the question has been around a lot longer than modern science, and for millennia, people have responded to this question in many ways. For some, the answer invokes either a moral or aesthetic necessity. In ancient Greece, for example, Plato believed that a constriction below the head was necessary to elevate the noble workings of the mind above the profane operation of the body. The head and the mind were supreme over the rest of the body. The gods even modeled the globe of the human head from the shape of the planets. They formed the neck for turning the head. But equally, the neck serves as an "an isthmus and a boundary" to keep the head apart and limit the flow of moral contamination from the lower mortal soul below.[5]

In one strand of Chassidic Judaism the neck was similarly conceptualized as a means of elevating the head above the body, but the constriction was viewed more as a moral tether or translator than a moral gate. Yes, the head is supreme and the seat of consciousness, but by itself it can be too ethereal and abstract. In this view, if the head were detached, it would be dangerously susceptible to "outside influences." The neck roots the head into our corporal foundation; it ties the aspirations of the head to "something that comes from within us and truly addresses our existence."[6] Thoughts are sparked in the brain, but as they are transmitted downward, the neck translates them into the earthly language of the body.

For the ancient Greek philosopher and natural historian Aristotle, the neck was a matter of geometry, not morality. He observed that the mouth is a single opening, but the paired lungs have two openings. So, logically, there needs to be a single pipe (the trachea) to carry air from the mouth to the two pipes (the bronchi) leading to the left and right lungs. Aristotle argued the lungs are what drive the necessity of the neck. Using reverse logic, he noted: "Therefore, it is that, when there is no lung, there is no neck." This truth was clearly borne out in the bodies of the fishes he dissected.[7] With these words Aristotle provided the explanation that was accepted, at least in the West, for nearly two millennia. Vesalius, the sixteenth-century Flemish physician often considered the founder of modern human anatomy, endorsed Aristotle's causal association between the neck and the lungs, titling one section of his treatise "The neck exists because of the lungs." But later, Vesalius emphasized that a crucial feature of the neck is to provide added space between the head and the chest for nerves to exit the spinal cord. The nerves supplying the torso already take up the exit points between the trunk (thoracic) vertebrae, so without a neck there would be no place for the nerves to the arms to exit. The neck (cervical) vertebrae thus provide this space.[8] For both Aristotle and Vesalius the neck was not the end product of a complex evolutionary history or a regulator of morality. It was simply a necessary linkage that enabled the splicing of tubes and cords from the head onto the organs of the body.

THE FUNCTION AND CONTENTS

Many people in contemporary times retain this view of the neck as merely a bridge joining two more essential body regions, the head and the torso. But this notion ignores the neck's many active, intrinsic functions. Indeed, the neck is the ultimate multitasker. It flexes, senses, vibrates, transports, and secretes every second of our lives. The conglomeration of functions located at the neck is one of its hallmark features and hardly equaled anywhere else in the body. As a foundation for appreciating the diverse activities and structures of the neck that are elaborated throughout this book, I begin with an overview of the basic physiology and anatomy of the neck, highlighting ways it differs from other body regions.

The variety of functions the neck performs occur at an especially wide range of speeds. Generally speaking, the systems of the body as a whole can be divided in two, based on the timescale of their operations. Sense organs and muscles work very quickly, on the timescale of milliseconds. They send and receive messages through electrical impulses that travel through the nervous system at more than 160 kilometers (100 miles) per hour. By contrast, digestion, excretion, immune responses, and reproduction work much more slowly, on the timescale of minutes to hours to weeks. These processes are controlled largely by chemical signals, such as hormones, that travel in the bloodstream at about 5 kilometers (3 miles) per hour. Most body regions specialize in one of these major functional divisions. Arms, hands, legs, and feet accomplish remarkably quick movements and fine tactile discrimination, but they contribute little to the slower maintenance and procreative projects of the body. The thorax and abdomen house most of the tubes, vats, and chemical factories that accomplish the long-term nourishment, physiological balance, and reproduction of our bodies, but they hardly move at all.

In contrast to these other body regions, the human neck participates vitally and equivalently in both functional divisions of the body.[9] We contract muscles in our necks thousands of times per day—about every six seconds—to move our heads in all directions, and we continually relay rapid messages to our brains about the position of our heads through sensors in our necks. We simultaneously pulse blood and inhale air every few seconds through the neck to nourish the body and brain. On a longer timescale, we ooze hormones from neck glands into the blood and throughout the body to regulate the pace of maintenance operations. Occasionally we deploy a platoon of immune cells through a vast network of lymph vessels. On top of these sensorimotor and maintenance functions, the neck accomplishes a nearly unique function: vocalization. The vibrations in the vocal tract are perhaps the fastest processes of all, occurring at hundreds of cycles per second. The neck simultaneously quivers and undulates with activity. These versatile functions of the neck are accomplished using a commensurate diversity of structures. Anatomically speaking, the neck has practically everything: bones that support, muscles that contract, cartilages that move and shield, cords that vibrate, tubes that transport, nerves that transmit electrical signals, glands that secrete hormones, and nodes that house immune cells. The neck's multifunctionality is enacted by a densely packed amalgam of parts.

The overall geography of the neck runs along several axes. The first axis is in the head-to-torso direction. In this axis the neck is basically a vertical column traversed by numerous tubes, cords, and straps that originate outside the neck. In general, these structures pass through the neck in one of two separate compartments, front and back, which form the second axis. The back half of the neck is mostly the domain of mechanics. The spine anchors the head, and long muscles attaching to the back of the skull and trunk stabilize and move the head. This portion of the neck is a pedestal, swivel, and motor. The front half of the neck, the throat, holds a diverse collection of tubes that carry all sorts of materials—gases, solids, and liquids—at a wide range of speeds. We pump waves of these materials through all these tubes many times every day, but the frequency of these pulsations differs considerably from tube to tube. The trachea, commonly called the windpipe, carries air back and forth between the face and the lungs about twenty thousand times a day. It has stiff ridges around its circumference, like a vacuum cleaner hose, to ensure that it remains open and always allows air to pass. Its sixteen to twenty cartilaginous rings hold it open at 2.5 centimeters (about 1 inch) in diameter, and these ridges give the trachea its anatomical name in Greek (*trakus*, meaning "rough").

The esophagus (in Greek, *oisophagos*, "that which carries and eats") transports slurries of food from the mouth to the stomach with every gulp; we swallow food or saliva about five hundred times a day. The esophagus is also about 2.5 centimeters (about 1 inch) in diameter, but in contrast to the trachea it is normally collapsed. Rather than resembling a vacuum cleaner hose, the esophagus is more like a toothpaste tube. It moves dense, viscous, chewed food downward by compressing the muscular walls of the tube. These two tubes run parallel through the neck with the trachea passing closer to the front of the throat and the esophagus more centrally. At the top end the esophagus and trachea join at the back of the mouth with a valve, the epiglottis, directing traffic. The intersection of these crucial tubes is the source of great vulnerability: choking. If the epiglottis fails to keep the food out of the trachea, we are thrown into a fierce fit of coughing or, worse, suffocation.

Running in pairs along the sides of these food and air transport tubes are large vessels for transporting blood. They pulse with freshly oxygenated blood about one hundred thousand times per day. The arteries have thick walls that can withstand the strong blood pressures necessary to supply the brain's stringent oxygen and nutrient demands. You can feel this pressure and its pulsation when you press lightly on the side of your neck. If you press too hard for too long, you will pass out because your brain does not receive enough blood. This phenomenon no doubt inspired Greek anatomists to name the main neck arteries the carotids, which comes from the verb *karotis*, meaning "to stupefy." To return blood back to the heart, jugular veins collect blood from the brain, face, and throat. Compared to the carotids, jugulars have lower pressure, but they are closer to the surface and have thinner walls, making them almost synonymous with vulnerability. Their name derives from the Latin *jugum*, meaning "yoke," and surely their vulnerability is a deep psychological yoke of the human condition. Our worst enemies and fiercest competitors "go for the jugular."

Entwined among these large blood vessels are the smaller, thinner tubes of the lymphatic system that play a crucial role in transporting cells of the immune system. Like thin strings joining widely spaced beads, lymph vessels connect a network of about three hundred cervical lymph nodes that house white blood cells ready to attack pathogens. Normally 1–10 millimeters (0.04–0.4 inches) in diameter, these lymph nodes commonly swell when the immune system is in full battle mode. These are the lumps doctors palpate when they check whether you are fighting an infection. By virtue of all these diverse pipes and tubes, much of the throat is about flow.

While these tubes transport chemicals and cells between the head and the torso, the many nerves of the neck electrically connect the brain to muscles, organs, and sensors all over the body. Near the core of the neck, the spinal cord, protected within vertebrae, is the primary electrical mainline in the body. Nerves exit the spinal cord between cervical vertebrae, and then, like railroads at a changing house, converge and diverge in a network termed the cervical plexus. Nerves passing through this plexus transmit touch sensations from the skin of the neck and scalp to the brain and activate muscles in the neck, tongue, shoulder, and diaphragm. Cranial nerves exiting directly from the brain pass into the neck to control muscles that move the head, open the jaws, constrict the throat, and operate the voice. One of these cranial nerves brings sensory information to the brain from a specialized organ in the neck, the carotid body, that detects pressure, oxygen, and carbon dioxide levels of the blood.

Aside from these many structures traversing the neck, several crucial structures are contained wholly within its borders. In the center of the neck there are seven cervical vertebrae and a few muscles that run between adjacent vertebrae. On the throat side the butterfly-shaped thyroid gland straddles the trachea and produces hormones that regulate the body's overall metabolic rate—the pace of our lives. The parathyroid gland sits on the surface of the thyroid gland as four pea-sized lumps that look like eyespots dotting the wings of the thyroid. Its secretions control blood calcium levels and thereby the composition of the skeleton—the scaffolding of our lives. Just below the jaw is the hyoid bone, which serves as an attachment for muscles of the tongue, jaw, and throat. The larynx, which juts out as the Adam's apple in many men, consists of four pieces of cartilage and connecting muscles that move the vocal cords in and out of the air flow to produce sound and up and down in the throat to modify the sound.

Tubes, pipes, valves, straps, cords, wires, sensors, bumps, and lumps. This long, diverse cast of actors perform their daily drama in darkness behind the curtain of the neck's thin and supple skin. Unless we look at an anatomical atlas, we are unaware of much of this complexity. However, one salient characteristic of the neck compared to other body parts is that its internal workings are far more perceptible at the surface. Usually the neck is lean and naked. Some of its muscles and sinews—especially the sternocleidomastoid, which passes diagonally from the back of the skull to the collarbone—protrude noticeably. With every swallow the hyoid bone bobs up and down visibly. Every breath and quiver of voice passing through the windpipe is audible. The warm pulses of blood, the vibrations of the voice box, and the swelling of our glands are all palpable to the touch. I believe that this semitransparency of its flesh and activity is at the root of our ambivalence toward the neck. Bare and exposed, it is both beguiling and disconcerting, alluring and just a little bit creepy. I suspect this is also why we lavish the neck with so much attention in adornment, attire, and erotica. The cultural drama on the outside plays in parallel with physiological drama on the inside.

• • •

What are the borders of the neck? For the purposes of this book, I define the upper boundary as the first cervical vertebra and the lower edge of the jaw and the lower boundary as the lowest cervical vertebra and the collarbone. There, the lines are drawn. But one of the interesting features of the neck is that in fact it has blurry edges. Its reach extends far beyond these borders. For example, certain muscle bundles inserting on cervical vertebrae extend as low as the pelvis. Certain nerves exiting the spinal cord in the neck reach all the way to the fingertips. Vibrations originating from the vocal cords in the throat pass through the mouth and as far as the forehead. Drawing the boundaries as above thus artificially severs the neck from some of its most interesting connections. So I occasionally spin out to anatomies beyond these borders either because they are naturally connected to the neck or because they tell such interesting stories that a diversion is merited for its own sake.

I need to confess another boundary, a taxonomic boundary. This book addresses only the necks of vertebrates: amphibians, reptiles, birds, and mammals. Many invertebrates, especially insects, have bona fide necks that function in many of the same ways as those of vertebrates. For instance, picture the highly mobile head of a praying mantis. As intriguing as they are, such necked invertebrates are beyond my expertise, and regrettably I have chosen to exclude them.

• • •

The upcoming tour of the vertebrate neck follows a rough anatomical direction, beginning at the core and radiating out to the surface of the neck. This direction tracks the basic chronology of the neck, both its formation in the embryo and its evolution among vertebrates. Generally speaking, early-developing and ancestral structures are centrally located while late-developing and more recently evolved structures are nearer to the surface. One can think of this core-to-surface direction as a third axis of the neck, one that parallels its history.

The first portion of the book begins at the core of the neck with the spinal column, which attaches the head to the body and supports its various postures and poses. Our longitudinal axis provided by the ver-

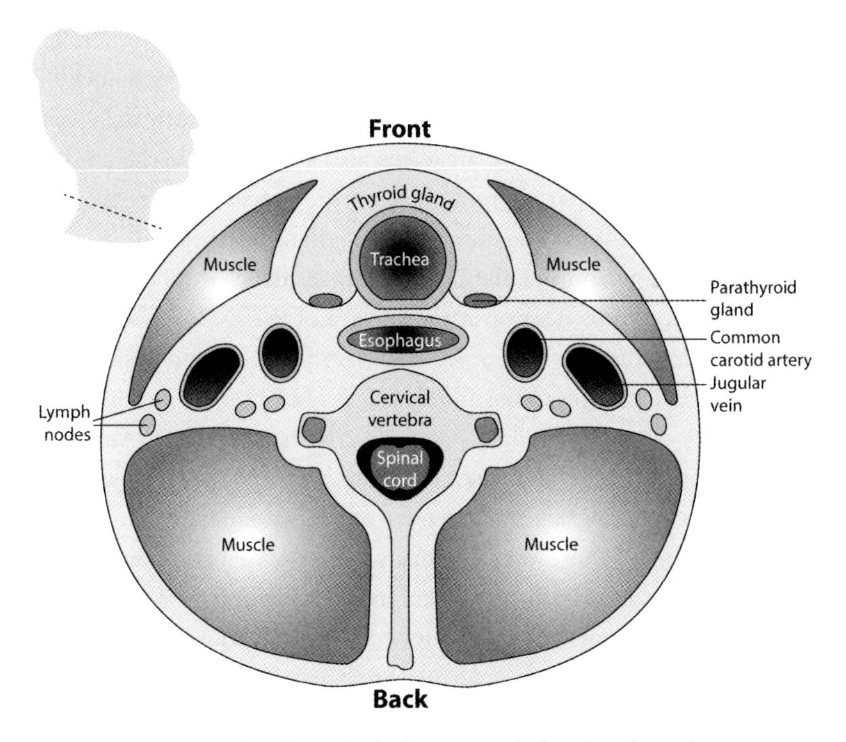

Figure 3. Cross-section through the human neck showing the major anatomical structures. Structures near the core of the neck generally arise earlier than those near the surface, both in the embryo and during evolution. Illustration by Netta Kasher, 2024.

tebral column is among the first features established in the embryo, and this ancestral feature unites us with all other vertebrates. Then we move to the muscles surrounding the core that move the head, giving us the capacity to scan the full panorama and to express a range of gestures through head movement. Muscles adjacent to the spine develop earlier embryologically and are more conserved evolutionarily compared to more superficial muscles. Next we examine the tubes and pipes—the esophagus, trachea, and blood vessels—that transport food, air, and blood between the head and the body. The esophagus, located more centrally, is part of the ancient gut tube that arises in the earliest embryological stages and is found in all vertebrates. The trachea lies more superficially and forms secondarily in the embryo as a branch off the esophagus. It appeared later in evolution, arising in terrestrial vertebrates that needed to move air to respire. Lying outside the trachea are the thyroid and parathyroid glands, which secrete hormones that regulate the pace of the body and the strength of the skeleton.

The middle portion of the book is largely devoted to how vertebrates use structures near or at the surface of the neck for vocal and visual communication. The voice box (larynx) is located visibly just under the skin. It arises in the middle stages of embryonic development and also appears in the middle stages of vertebrate evolution. As vertebrates colonized land and began breathing by transporting air into lungs, they could use this air movement to generate sounds for communication. Many of these vocalizations serve in sexual communication. During courtship, animals often combine these mating calls with alluring visual ornaments on their surfaces—modifications of the skin, plumage, and fur such as colorful throat patches and fluffy manes. Sexually distinct voices and courtship ornaments usually appear late in development at the onset of sexual maturity. Humans also use the neck in sexual communication, with both the voice and neck ornamentation—jewelry, perfumes, and neckwear—all contributing to the intricate dance of courtship. Many of these same surface ornaments are used to communicate status and identity in both animal and human societies.

The final portion of the book examines how humans and animals exploit and guard against the vulnerability of the neck. Most of this violence and defense occurs at or near the surface of the neck. Carnivores, both animal and human, often kill at the neck by piercing or slashing the skin. Humans control each other and our domestic animals with devices that encircle the neck—shackles, nooses, and yokes. But animals and humans have evolved and invented means to protect and soothe themselves from the neck's vulnerabilities. Lymph nodes just under the skin help fight infections; massage therapists pressing into the skin relieve aching neck muscles; armor and amulets worn on top of the skin guard against all sorts of threats. Finally, in the epilogue, I consider necks that are products of our hands rather than evolution. Humans, unbound by biological constraints, form ceramic vases whose necks must sometimes serve specific functions but that mostly serve to please our eyes. I consider what these inanimate necks might reveal about the nature of living necks.

With this map laid out, let us begin with the simplest, most ancestral of all neck functions: holding up the head.

CHAPTER TWO

Posture & Pose

Holding the Head

At a glance, the people walking past us on the sidewalk display many of the hallmark features commonly associated with the human species. They stride down the sidewalk bipedally while their opposable thumbs dash around on a smartphone, a remarkable achievement of their oversized brain. Much of their thumbing produces abstract communication signals—texts—in the service of complex social interaction. They are *Homo digitalis*: bipedal, manual, smart, verbal, and connected. They enact their social life with agile digits sending digital messages.

As members of *Homo digitalis*, we commonly flex our heads forward while manipulating our phones in our hands or viewing our tablets flat on a table. Some medical professionals have claimed that this device-induced posture is leading us into an epidemic of cervical pain and malformation; others have retorted that humans have always hunched over what they attend to at close range—a stone tool, a baby, or a book—and that devices are unfairly blamed. The cervical malady known as "text neck" was given its name in 2008 by chiropractor Dean Fishman, who eventually trademarked the term. Six years later, text neck splashed on the mainstream media, following the publication of an article by spine surgeon Kenneth Hansraj, who calculated the forces exerted on the cervical spine by the head tilted forward at different

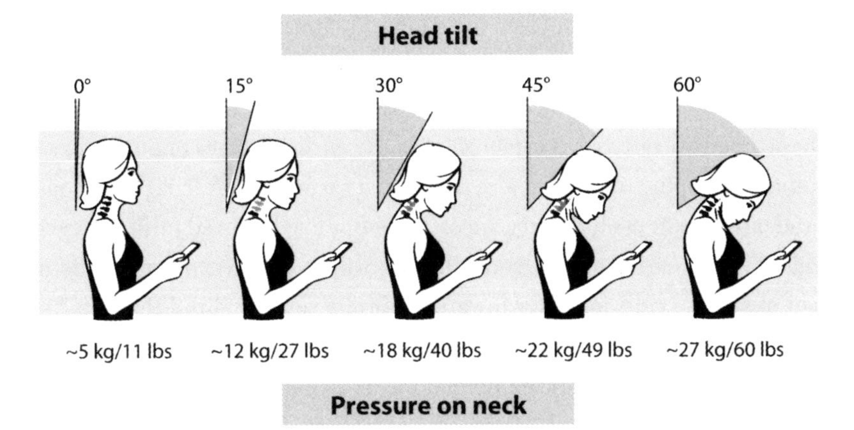

Figure 4. The effect of operating a smartphone with different head angles on the force imparted to the neck. From Hansraj 2014, 278.

angles.[1] When the ears are directly above the shoulders and the neck is in the so-called neutral position, the head exerts about 5 kilograms (11 pounds) of force. As the head tilts forward into more typical device-using postures, the force increases to 18–27 kilograms (40–60 pounds), about the weight of a seven-year-old child. The internet ignited with warnings and therapies but also with rebuttals.

Within three years more than 9,900 scientific articles were published on this twenty-first-century dilemma. Systematic reviews of the scientific literature concluded that *Homo digitalis* does indeed face unprecedented threats (although far from epidemic) to its cervical well-being.[2] Frequent cell phone and tablet usage tends to cause greater neck flexion and increases the prevalence of neck and shoulder symptoms. Using these devices is probably different from other close-focus activities because we tend to hold cell phones relatively low and brace our arms against our bodies while thumb-typing and simply because we use them for so many hours of the day. Figuring out the full consequences of this posture will require more study, reading, and writing, which no doubt will be done with tilted necks and heads bent over screens.

• • •

Perhaps the neck's most basic function is to attach the head to the body and hold it in place. It does so largely through modifications at one end of the long axis, the spine, that extends from the head to the tail. This basic head-to-tail axis is established in the earliest stages of our embryonic development (in humans, at about two weeks after fertilization) and our distant evolutionary ancestry (about five hundred million years ago).[3] In the early embryo, soon after it forms a hollow cluster of cells, a set of surface cells migrates inward to form a semirigid rod that sets up the longitudinal orientation of the animal.

This early structure (the notochord) is so fundamental to the body design that it served as the namesake of our entire phylum, Chordata. A bit later in development the notochord forms the armature onto which other skeletal and muscular structures are laid out. On either side of the notochord a set of tissue blocks (somites) condenses and progressively forms the series of bones (the vertebrae) that defines our class of backboned animals, Vertebrata. In most vertebrates the notochord mostly disappears in the embryo, but it is retained as part of the spongy cushions (the vertebral discs) between the vertebrae. The vertebrae are eventually linked together with muscles and ligaments, giving vertebrates a certain stiffness and directionality from head to tail. Because the spine is narrow and composed of a series of mobile elements, the body can also bend and torque. As a connector for the head, the cervical spine must serve both these basic mechanical functions: stability and flexibility.

In our fish ancestors the spine held the head horizontally, which was crucial for aquatic locomotion in the head-to-tail direction, but because the water supported their body weight, they did not need robust skeletal or muscular structures to hold the head. This basic horizontal posture was retained in most necked vertebrates—quadrupedal, terrestrial animals such as most amphibians, reptiles, and mammals. However, on land, vertebrates needed more sophisticated means to counteract the force of gravity, which tended to bend the head downward. Terrestrial vertebrates also commonly angled their necks upward a bit to extend their range of vision. So, many of them evolved a complex series of joints, muscles, and tendons that keep the neck stable but flexible and in a slightly tilted but mostly horizontal position.

But, in some cases, including many birds and a few bipedal mammals such as humans, the geometry of head support totally changed. In human evolution heads became upright, balanced on a vertical vertebral pedestal, and gravity tended to compress rather than bend the cervical spine. With the head supported mostly underneath, human necks and those of other bipedal animals can be slighter. However, with a less muscular, less sinewy neck, humans are prone to chronic strains when we lean our heads forward for long periods and susceptible to neck injuries when sudden, unpredicted forces act from the side. Our bipedal posture, which had so many evolutionary advantages in elevating our senses and freeing our hands, thus makes us vulnerable to modern, technology-driven threats such as text neck from our devices and whiplash from our vehicles.

In every moment of the day, our head/neck posture must meet the demand of mechanics—withstanding or counteracting external physical forces—but the particular way we hold our heads also signals in a single snapshot much about our interior world. Seen by others, posture is a pose, expressing mood and attitude. The head bowed downward shows sorrow or submission; the head raised with a little twist conveys scorn or haughtiness; the head rocked left or right signals curiosity or uncertainty. By controlling the position of the head, the neck has potent communicative and expressive capacities. We all commonly attend to these head positions in reading the nonverbal body language of others. Below, I elaborate on this duality in the way humans and animals use their necks to hold their heads: the mechanical vulnerability and the expressive vitality.

POSTURE: MECHANICAL CAPACITIES AND VULNERABILITIES

Humans have an (almost) perfectly balanced head and a neck that is well-designed to hold it in an upright posture. Our closest living relatives, the chimps, have a more forward-tilting spine and require much more effort to keep their heads from flopping forward. So even though a human head is proportionately the same size as a chimp's head (about 8 percent of body weight), we require only about one-tenth the muscular energy to hold our heads.[4] The greater balance of the human skull (compared to our ape relatives) is also attributable to the foreshortening of the human face.[5] With our small "muzzle" we do not hang a lot of weight out in front of the spine/skull joint. Moreover, the weight of our face in the front is counterbalanced by the rearward bulge of the brain case behind the spine/skull joint. Nonetheless, the center of mass in our skull is still slightly—about 1 centimeter (0.4 inches)—forward of the joint, and without any mechanical support our head would rock forward.

Fortunately, much of this support comes from the highly elastic nuchal ligament—an anatomical bungee cord—that spans between the vertebrae and back of the skull. This elastic support combined with the vertical alignment of the head atop the cervical spine means that humans, compared to other primates, have small neck muscles for stabilizing the head. Leonardo da Vinci seems to have been very aware of the elegant design of our vertical neck in supporting the head. In his anatomical sketchbook only a handful of the more than two hundred anatomical drawings are accompanied by mechanical or architectural sketches. But on one of the plates with his drawing of the upright neck, he drew an adjacent sketch of an architectural column, capped at the top with a capital, linking parallel designs in anatomy, art, and engineering.[6]

Superficially, the balance of the head on the spine seems like a clear design advantage and one that evolved as an adaptation for our upright stance. But, as Daniel Lieberman explains in *The Evolution of the Human Head*, a closer look shows that that advantage is more equivocal.[7] For four million years our hominin ancestors, including Australopithecines, traversed wide expanses walking bipedally, but they had very unbalanced heads. Moreover, other mammals that are bipedal (e.g., kangaroos) or otherwise have mostly vertical necks (e.g., giraffes) also have unbalanced heads. There may even be advantages to an unbalanced head that have promoted its persistence in species with erect necks. One of these advantages is in stabilizing the head during locomotion. When walking, all animals encounter the problem of head instability. Quadrupeds (most amphibians, reptiles, and mammals) have their heads cantilevered forward from a spine, and many contract the muscles on the back of their neck in rhythm with their stride to offset the undulating motions of the body to stabilize the head.

Since the human head rests directly on top of the spine, we cannot use this counterswaying method, and head stabilization during locomotion becomes problematic. When humans walk, the head, connected by the spine to swinging legs, continually bobs up and down, about 4–5 centimeters (1.5–2 inches) with each stride.[8] With this vertical bobbing, we face what Lieberman calls the "pogo stick" problem. Every step jolts the head upward. The problem of head instability gets even worse when we run. Each stride inflicts an upward pogo stick force that is about three times the weight of the body and bounces the head upward about 10 centimeters (4 inches). When running, we switch our support from leg to leg, and the body sways left and right with each stride. The combination of these up-down, left-right forces can be seen rather playfully on the heads of joggers: ponytails bobbing in a figure eight displaying the complex mechanical forces acting on the head of a runner. To counteract this pogo stick problem, we use several mechanisms to limit the jostling of the head during running.[9] The left-right sway is buffered by forward swinging of the arm opposite the foot that contacts the ground. Rhythmic contractions of the muscle on the back of the neck (the trapezius) reduce the forward roll of the head that would otherwise occur because the torso leans a bit forward when we run. With these two compensatory movements, we can minimize jostling our head when we run.

Beyond the pogo stick problem, an additional disadvantage of our upright posture is that it crowds the throat. Picture an angle with one side from the chin to the back of the jaw and the other side from the back of the jaw to the collarbone. In quadrupedal animals—think about the neck of a dog—this is a broad angle. But in humans the angle is much narrower because the head is, by comparison, tucked down toward the torso. This compresses the space for all the organs in the throat, which, as the anatomist Vesalius noted centuries ago, is already the most crowded region of the human body.

A further space-crunching consequence of human bipedality is that the neck is short compared to most quadrupeds. For quadrupeds the distance from the chin to the base of the neck has to be at least as long as the legs for them to nibble objects on the ground. In their quadrupedal posture they have plenty of space to suspend organs from their long and more horizontally positioned cervical spine. For humans, however, it's a different story. With hands as our primary means of grasping objects, we are free to have shorter necks. This shorter neck helps with head stability but constrains the dimensions of such organs as the trachea, larynx, thyroid gland, and throat musculature. If all our neck had to do was to support its capital, it could be as solid, long, and elegant as an architectural column. But the human neck is cluttered with structures that serve other vital functions, all of which require space and must withstand a lot of bending and bouncing.

• • •

In our bipedal posture gravity pushes (mostly) straight down on the head and spine, relieving the neck muscles from great effort in stabilizing the head. This posture enables humans to accomplish a remarkable task unknown among other creatures: we can carry massive objects on top of our heads, with balance sensed and partially maintained through the neck. Other animals might drag around large prey in their jaws, some mammal parents (including some cats, rodents, and primates) carry their babies in their jaws, and elephants carry logs in their trunks, but no other animal balances their cargo on top of their heads. Across the world people have used their heads to carry all sorts of loads, including water, firewood, bags of grain, and even construction materials.

In one viral YouTube video a young man from Bangladesh stands on a small boat filled with bricks.[10] His coworker hands him two bricks that he places on his head with his hands. Two more. Four, six, and on and on. At fourteen bricks, the pile on his head exceeds the length of his arms, and he begins to toss bricks, two by two, where they land perfectly aligned on the pile, which is perfectly aligned on his vertebral column. As the stack grows taller, each toss requires a little thrust from the legs. At twenty-two bricks, a stack that weighed about 40 kilograms (roughly 88 pounds) and rose about two-thirds his height, the porter turns and, with barely any caution or break in stride, walks a 30-centimeter-wide (1-foot-wide) plank off the boat onto the dock. He cannot see the stack and balances it by feel, largely through his neck. In this video the porter transports his load a short distance to the dock, but historically head porters have faced much longer, arduous journeys. In one epic episode porters carried pieces of large ships on their heads hundreds of miles, all in the service of European colonization of Africa.[11]

In the late nineteenth century the British became quite interested in controlling the interior of East Africa, the Protectorate of Uganda in particular. This region gave them access to the headwaters of the Nile River, which was strategically important for securing all of northeastern Africa. One key to controlling this area was naval and shipping dominance of the large lakes in the interior, Lakes Victoria and Tanganyika. But there was a problem: how to get steamships built in Britain to these lakes that are far from the coast. Their solution included dismantling the ships into parts and loading the packages on the heads of African porters, who carried them in caravans more than 900 kilometers (almost 560 miles) through arid deserts, dense, parasite-infested jungles, and rugged mountains to the shores of the inland lakes of the Rift Valley.[12] One ship transported by head porterage was the SS *William Mackinnon*. In 1890 the steamship was constructed in the shipyards of Scotland and then disassembled into over three thousand packages, most weighing about 27 kilograms (60 pounds), the ideal weight for human porterage. The parts were shipped to the port city of Mombasa (now part of Kenya), and for several years the packages were transported westward by foot along paths that were part of an extensive network in East Africa. Eight years later, the parts made it to the shores of Lake Victoria, where they were assembled into a functional steamship. The SS *William Mackinnon* served continuously until 1929, when it was "condemned as unfit for further use" and sunk to the bottom of the lake, burying the labor of thousands of head porters.[13]

Head porterage continues in many parts of the world, and as means of human-powered transport, it can be surprisingly efficient ergonomically.[14] However, it is increasingly clear that headloading exacts enduring costs on the skeletal components of the neck.[15] When weight presses down on the neck, it mostly exerts compressive forces on the cervical spine.[16] Bone resists such forces well, but the soft structures between the bones (the intervertebral discs) are far more vulnerable. These discs, of "slipped disc" notoriety, are mostly cartilage, and they cushion and provide flexibility at the junction between vertebrae. The center of each disc is filled with a gel-like substance that degrades with normal aging. Together, these discs comprise almost one-fifth of the spinal column length, and their age-related deterioration contributes to routine decrease in height in the elderly.

Of greater consequence, when the cushion between the vertebrae decreases as we age, the bones tend to rub against each other, leading to arthritis, or pinch the nerves that exit the spinal cord between the vertebrae, leading to pain or numbness in the limbs. When porters place heavy loads on their heads, the thickness of the cervical discs immediately flattens, and over time the discs commonly degenerate in a condition termed spondylosis. In numerous studies head porters show incidence of cervical spondylosis three to four times greater than control subjects working in other labors. Not surprisingly, head porters also show more chronic neck disease and pain.[17] So, while headloading has enabled people to transport massive quantities of cargo by foot, head porters still pay the price in long-term suffering at the neck.

• • •

One peculiarity of the way that we hold our head is that our neck (and the entire spine) is oriented perpendicular to the direction in which we normally locomote. So, if we collide into objects—or more important, if they collide into us—the horizontal forces tend to whip the neck back and forth in extension and flexion. Because of our relatively weak neck muscles, we have little capacity to resist these forces. Our vertical necks therefore make us vulnerable to whiplash. For most of human history this vertical design was seldom a problem for our locomotion. As we moved around the world slowly, propelled by our own legs or those of domestic animals, the forces during collisions were only rarely strong enough to whip the neck. However, beginning in the early nineteenth century, trains and cars began zipping us around at much faster speeds. With this advancement we discovered whiplash as a new neck vulnerability. But wait, did we discover it or invent it? That's the controversy. It's a complicated, high-stakes debate. "There are two great puzzles in this world that foster debate among humans. One is the wonder of the universe, the other is whiplash," writes Dr. Murray Allen, a prominent Canadian researcher and physician.[18]

Few would argue about whether car collisions can cause short-term neck pain. They do. The debate surrounds chronic pain, the kind that can debilitate a person for years and lead to protracted legal action and hefty exchanges of money. The crux of the debate is that whiplash is always subjective. Neck pain following collisions is undoubtedly real, sometimes brutally real, but it is a reported experience, not an objectively observable phenomenon. Certain neck disorders such as arthritis, slipped discs, or tumors can be identified with little ambiguity and without any input from the patient. But so far, no physician can look at a medical image or take any measurement and confidently report that a person definitely has whiplash. Rather, whiplash can only be identified from a person answering the question "Does this hurt?" Sometimes, especially in the courts and insurance offices, that is a multimillion-dollar question. Americans make insurance claims for neck sprains more than any other injury.[19] The claims for whiplash injury total more than $200 billion annually.[20]

On one level the cause of whiplash is simple: a rear-end collision thrusts the thorax forward more forcefully than the head, and the neck rapidly hyperextends backward. As the car decelerates, the head lurches forward, but because the thorax is heavier and restrained by a seatbelt, the neck recoils in hyperflexion.[21] All this in less than a quarter of a second. Ouch. The neck vertebrae are not typically damaged, but muscles and ligaments spanning between bones are commonly stretched and damaged. Such sprains likely account for much of the initial localized pain. For most people these initial symptoms resolve soon after the accident. However, for about 20 to 25 percent of folks the symptoms persist for months or years and commonly extend into other body regions: pain in the back and limbs as well as the neck, headache, jaw disorders, sensory (vision, hearing, and balance) disturbance, and cognitive impairment.[22] Whiplash symptoms also vary considerably among individuals. For example, older people, those with preexisting cervical disorders, and folks with previous psychological distress are particularly prone to chronic whiplash symptoms. In addition, a patient's attitude toward the injury makes a difference. Patients who expect to recover quickly from whiplash in the end usually recover more quickly than patients anticipating chronic pain.[23]

Cultures report widely different experiences of whiplash. For instance, in the United States, Canada, the UK, Ireland, and Scandinavia, people who experience acute whiplash are far more likely to report chronic symptoms than people in Germany, Greece, and Lithuania, even when they have experienced similar car accidents. In some countries chronic whiplash is not even recognized as a category of injury.[24] Robert Ferrari, author of *The Whiplash Encyclopedia*, cites studies suggesting that the culture of litigation may contribute to national differences in the epidemiology of whiplash.[25] For example, in the early 1980s Australia had a relatively lenient legal procedure for seeking financial compensation for auto injuries and correspondingly had a high rate of complaints of chronic whiplash. In nearby Singapore, where such legal options were less available, whiplash was rarely reported, even though the rate of auto accidents was similar. When Australia instituted legal changes in the late 1980s that tightened the conditions under which an injured person could file for compensation, the reported incidence of whiplash fell precipitously. However, the cultural differences are likely due to more than the legal systems. Ferrari concludes that even beyond the possibility of financial compensation, cultures that have adopted the expectation that acute neck injury leads to chronic pain are more likely to report higher rates of chronic whiplash.

Still, the field of whiplash remains full of controversy. Those contending that whiplash arises foremost from tissue damage rather than psycho-cultural interpretations of injury insist that future diagnostic tools will be able to identify unambiguously the material cause of whiplash pain.[26] But for now it's unclear whether medical technologies will ever enable us to fully disentangle the complexities of injury, pain, and belief that can follow the horror of car collisions. In the meantime our automotive technologies propel us through the world at ever faster rates and expose us to forces that far exceed those that shaped our bodies and posture over evolutionary time.

• • •

Whiplash is a risk arising from our weak and upright neck, and most of us are willing to take this risk for the sake of quickly moving around the world horizontally. A select few are also willing to accept substantially greater risk to the neck for the thrill of moving vertically. For sport and excitement, these people choose to launch themselves from cliffs up to 30 meters (100 feet) high and plunge, sometimes headfirst, into the glittering water below. Fatal, upside-down whiplash is one miscalculation away. If the cliff divers do it correctly, they hit the water straight on and their cervical spine absorbs the compressive forces of the water, but if their entry is not directly head-on, the bending forces can immediately cause serious injury, paralysis, or even death. Amateur recreational diving is the greatest source of recreation-related quadriplegia and the fourth overall cause of hospital admission for spinal cord injuries in the United States, trailing only car/motorcycle wrecks, falls, and gunshot wounds.[27] Trained professional divers fare better, but still, in one survey of cliff divers, 14 percent of them suffered severe injuries in a single year, with the cervical spine and the neck as the second most common injury site.[28]

Despite the dangers of high diving, I understand the allure. I think one of the most thrilling of all spectacles in nature is the sight of seabirds dive-bombing for fish into the sparkling sea. A cloud of gannets swarms 30 meters (100 feet) above the water, each one bending its head downward to scan for sardines and occasionally rearing up in a partial hover. One by one in a syncopated rhythm, they roll into a vertical position, stiffen their necks, and plunge downward with kamikaze intention and abandon. Just before hitting the water, they point their wings backward and then bullet their bodies 10 meters (30 feet) underwater to snag their meal. An individual gannet makes as many as a hundred such dives in a single foraging trip. Yet, unlike kamikaze pilots or cliff divers, gannets almost never die or get injured. Perhaps that is why it is so thrilling to watch: you know there is an improbable happy ending. Considering the physics of the gannet dive, this happy ending is not a forgone conclusion. The head of a diving gannet, supported by a neck that is only about 5 centimeters (2 inches) in diameter, hits the water at more than 100 kilometers (60 miles) per hour. To make matters worse, the water surface might also be churning erratically in a way that makes a clean entry unlikely. Surprisingly, the initial collision with the water is not even the most dangerous moment. Just after the head enters, it momentarily opens up a gap of air around the neck and in this phase the neck is most susceptible to irregular forces that could cause it to buckle and break.

Brian Chang, a biomedical engineer at Virginia Tech University, and his collaborators investigated the physics of seabird diving to find out why such dives are surprisingly safe for the birds.[29] They first examined the fluid dynamics of the dive by taking a dead gannet, freezing it into its arrowlike diving position, and plunging it into water at velocities that replicate a real dive. Then, to simplify matters, they 3D-printed a plastic cone that roughly mimicked the shape of a gannet's long-beaked head and attached it to a plastic "neck" that had mechanical properties similar to those of a real gannet neck. With this basic model they could vary the anatomical parameters and dive speeds to see which properties made the artificial neck prone to buckling when the model was dropped into water. With certain artificial anatomies—for example, broad beaks, long necks—the neck indeed failed. But with shapes approximating a real gannet, there was never a risk of instability.[30] Moreover, when the researchers incorporated the effect of neck muscle contractions that actively stiffen the neck, the safety factor was even greater: the force required to buckle such a contracted neck was about a hundred times greater than what the bird normally experiences during a dive. In fact, they estimated that a real gannet, with this level of "overbuilt design," could withstand a dive of almost 300 kilometers (180 miles) per hour, three times their typical dive speed. Remarkably, it looks like gannets with their high-speed vertical dives are actually playing it exceedingly safe.

Here on land, moving slowly and horizontally, we humans devote little precision, effort, or complex anatomy to holding our heads. In our bipedal stance we are usually blessed with balance. But our thin, upright necks also make us vulnerable. We sometimes must endure neck pain as a side effect of distorted posture in our technology consumption. And we can never fully escape the fear that unexpected forces will cause us to slip a disc or whiplash our necks. Our vulnerability is the flipside of our balance.

POSE: EXPRESSIVE VITALITY

The way humans and animals hold their heads enables them to have both remarkable capacities (e.g., head porterage and plunge-dive foraging) and deep vulnerabilities (e.g., text neck and whiplash). But, in addition to providing the structural support to withstand these physical forces, the neck has a more nuanced role in both daily life and art. It helps convey emotions and thought. It communicates through poses. Animals as well as humans communicate through head posture, and this is most familiar in domestic animals. Dogs lower their head in submission or playfulness; they raise it in aggression or fear. Horses with a lowered head are generally relaxed, but if they sway their neck laterally at the same time, it is an act of aggression. When they elevate their head, they signal alertness and perhaps worry about a distant threat, but when they rear their head while being ridden, they are likely in pain.

From our daily experiences with humans as well as animals, we are continually reminded of the role of head position in expressing emotion. But for Aristotle, the angle of the head also revealed contemplation.[31] For him, it was not coincidental that people lost in thought hold their head tilted backward; this angle helps you think. He asked, "Why doth a man lift up his head towards the heavens when he doth imagine?" Aristotle believed that the powers of imagination resided in the front of the brain. When the head is lifted in thought, "the spirits which help the imagination" rise to stimulate the forebrain. He asked, "Why doth a man, when he museth or thinketh of things past, look towards the earth?" The back of the brain houses the powers of memory, he thought, so that when you hold your head tilted downward, "the spirits which perfect the memory" can flow upward into it. This is not exactly how moderns conceive of cognition, but you have to admire Aristotle's combination of human observation and logical inference. And, if thoughts were indeed stimulated by particular head positions, Aristotle's head, filled with so many thoughts, must have been in constant motion.

• • •

Evolution has endowed humans with an anatomy and posture that requires little effort to hold our head in the straight-forward, neutral position. But apparently we do not like to look at ourselves in this position. Most people hate their driver's license photo or any other bureaucratic portrait, such as a passport photo or a mugshot. These photos function in individual identification, but ironically at the same time they erase our individual personalities and expressions. With our heads square on our shoulders, looking neither up nor down, left nor right, we are blank and neutralized. Fortunately, we seldom have to look at such lifeless photos. Flip through a magazine, walk into an art gallery, or survey the advertisements that bombard us daily, and you will find that head-on, frontal poses are rare indeed. In most portraits, fashion photos, campaign posters, theater headshots, or any composed depiction of a person, the neck holds the head in an angled and expressive manner. Evidently, one component of humanizing and personalizing a head is to give it a tilt or a twist. While such expressiveness may emanate from the head, these poses are actually an action of the neck; heads are held in position or moved only through neck muscles acting across neck joints.

It is not just models and celebrities we prefer with angled heads; we like *ourselves* that way too. Ours is the golden age of the self-portrait: the selfie. In many beautiful and popular spaces, from mountaintops to parties, you commonly find someone with an outstretched arm and a tilted head. Either deliberately or subconsciously, we rock our heads in all three orientations to get that just-right angle and lighting. As spontaneous amateur self-portraits, selfies have captured the attention of psychologists interested in the expressive nature of the head and neck. Are there any patterns to these intuitive, just-so poses? Using a database of more than three thousand selfies posted to Instagram in six global cities, researchers found that head position in selfies is far from random.[32] Confirming our apparent loathing of the mugshot pose, only about 7 percent of selfies are full-frontal views. Among those with turned heads, there is a clear bias toward turning the head to display the left cheek. When independent viewers were asked to rate the attractiveness or "emotional intensity" of the portraits, they rated left-displaying faces higher than frontal or right-displaying faces.

Psychologists have long known that people respond more emotionally to images of the left side of the face, and this may arise from left-right differences in the brain.

Another familiar head adjustment in selfies is tilting the head upward in the vertical direction. In their typical poses selfie-photographers often stretch their arm high above the head to show their head tilted upward. This posture creates a "height-weight illusion"; it makes us appear thinner.[33] The vertical tilt of the head apparently plays a particular role in courtship communication. When researchers examined selfies on the dating app Tinder, they found that women show a strong tendency to look up at the camera while men more often look down. The authors of this study speculate that this bias relates to sexual differences in height. Among heterosexual couples, women more commonly look up at their romantic partner, and men commonly look down. So courtship selfies may prospectively enact this couple's posture.[34]

• • •

A walk through an art museum shows that our predilection toward bent or turned heads far predates selfies or, for that matter, photographs of any sort. In Western art, beginning in classical Greece, the vast majority of portraits in painting and sculpture depict their subject with an angled head. When researchers compiled data from the entire collection of paintings, drawings, and photographs in London's National Portrait Gallery and portraits of more than four thousand "historically significant" Americans, they found that fewer than one in seven of the sitters had heads with full-frontal presentation. In most cases the whole body was turned askew, with the head turned back to gaze at the viewer (think *Mona Lisa*) or with heads and bodies oriented in the same direction and the sitter gazing at an angle into the distance.[35] This study and many previous surveys showed that, just like untrained selfie portrait artists, the portrait masters in fine art tend to display the left side of the face. So artists—or perhaps the sitters themselves—may turn the head of their subjects to elicit a strong emotive response from viewers.

In addition to this rotation of the head, many paintings—about half in one survey—display the head "canted"—tilted in the left-right direction. Religious figures, which were commonly projecting emotions such as pity, mercy, or adoration, were far more likely depicted with a canted head than more stern authoritarian figures, such as royalty and nobility. Notably, for both skewed postures (left-bias turns and sideways tilts), women, who are more commonly associated with outward expressions of emotion, were more often depicted with an angled head than men.[36] Altogether, a tour of European and American portrait galleries confirms what we suspect about the lifelessness of our driver's license photos: an angled head enhances the expressive potency of the portrait.

Beyond the formal portraits of individuals, most Western paintings with groups of people depict their subjects with angled heads. In some cases, where the subjects bend their heads toward the same focal point, the effect on the viewer is to concentrate attention on an event or object. The viewer knows what is important because everyone in the painting is focused there. More often, the multiple subjects in a painting have heads turned in different directions. One effect is to convey divergent emotive responses to the same event. In Fra Angelico's fifteenth-century painting of the Crucifixion, for instance, one kneeling woman looks up at Christ in desperation, another standing woman bows her head in prayer, one man holds his head askew in astonishment, another looks down to read a book, and some seem to look around at each other to gauge their reactions. In other paintings, angling the heads of the subjects in multiple directions gives the sense of widespread movement and activity. Rembrandt's *Nightwatch* depicts the gathering of a civic militia company in Amsterdam, complete with drums, guns, and flags. It almost trembles with commotion. Thirty-four faces appear in the painting; almost all of them are looking in a different direction as indicated by the angle of their head, heightening the clamor of activity. As viewers, we use the position of the head to assess what subjects attend to and care about as well as their emotive responses.

Figure 5. (a) Sculpture of Saint Sebastian by Gian Lorenzo Bernini, 1617. Museo Thyssen-Bornemisza.

Figure 5 *(Continued)*. (b) Sculpture of the Buddha, Afghanistan, ca. 300 CE. Cleveland Museum of Art.

In Western art there are some notable exceptions to the general pattern of turned heads. Figures of strength and authority—those impervious to the vagaries of emotions—commonly look straight forward. In the dozens of portraits of Henry VIII, for example, the king looks squarely at you, with his head and thick neck oriented squarely on his shoulders. In military recruiting posters, the model soldier is similarly depicted with a broad neck holding the head sternly forward. It's not just men. Queens sometimes orient straight toward the viewer (e.g., the Coronation portrait of Elizabeth I [ca. 1600]). The blind lady of justice, with a sword in one hand and scales in the other, looks straight on. Lady Liberty gazes out at New York and the world with a straight, unswerving neck and strong-willed stare. Lifeless people also have straight necks. The honored dead, usually royalty or clergy sculpted as effigies, often lie flat with their head square on their shoulders looking straight up. They rest in peace; death has released them from their emotional reactions to this world. It's more peaceful to rest eternally with the head in the neutral position.

In some works the artist appears to intentionally replicate the expressionless posture of a mugshot. In the iconic *American Gothic*, painted in rural Iowa during the depths of the Great Depression, Grant Wood portrays a stern farmer and his daughter with near documentary detachment. The commentary of the Art Institute of Chicago, which houses the painting, notes that the "rigid frontality" of the portrait is one of its defining features. The farmer, in his blank stare and fully frontal posture, seems solid and stoic while the daughter, though still stoic, appears a bit more worried. With a tense frown and a slight twist of the neck, she seems to peer anxiously beyond the artist to an unknown future.

Keep walking through the art museum and when you get to the wings displaying pre-Hellenic Western art and non-Western art of many periods, you will find that heads in the neutral position are more the rule than the exception. Indeed, considering art across the full scope of human history and cultures, the Western, modern preoccupation with neck torsion and tilt is a brief and provincial inclination. In almost all art of precolonial Latin America, Africa, and Oceania (at least that which is displayed in museums), figures are depicted with heads square on the shoulders. The art of Asia is more variable, but still, the majority of figures seem to have straight necks.

One posture that appears to connect deeply to a balanced emotional world is the neutral neck and balanced head of the Buddha. The Buddha is likely one of the most sculpted and painted figures in human history. As far as I can tell, he almost never bends his neck. In both sculpture and painting, the Buddha usually sits with his legs crossed underneath him, his arms in his lap, his eyelids slightly open, and his head straight forward. He favors neither left nor right, up nor down. He is balanced. Rather than conveying emotional blankness, his frontal posture signals the strength of emotional stability and disengagement. As a young man, the Buddha traveled widely and experienced the turmoil of human suffering. But, in his path to enlightenment, he sat still under a Bodhi tree, neck straight in the yogic posture, learning to detach from whatever suffering or external temptation might pass by. By the time he arrived in this contemplative posture, he had already had all necessary interactions with the world, and he could move toward the final spiritual transformation. He could see all he needed by paying attention to what was directly in front of him. There, the Buddha sits: aligned, frontal, indeed neutral, but in a composed, serene way.

• • •

I end this chapter the way most of us end our day: with sleep. The columnar neck serves us quite well as we pass through the day in vertical posture. But when we get horizontal to sleep, the head cocks or twists to one side, and we are apt to wake with a crick in the neck. When we try to lie on our bellies, the problem is the orientation of the head necessitated by our bipedal design. Unlike quadrupedal animals, we cannot lie on our front and rest our chin on the ground. We have to torque our head 90 degrees to the side. When we lie on our side, the problem is our broad shoulders. In contrast to quadrupedal mammals, humans are compressed front-to-back, and our shoulders extend far laterally to our head. So, lying on our side, our head flops down, tilting our cervical spine. When most mammals lie on their sides to sleep, their heads are not far from the ground, and their necks are mostly aligned with their spine. In species that sleep in groups (e.g., many primates, dogs, and cats), they may support their heads on each other in sleeping conglomerations. Humans also commonly lay their heads on each other or on their own arms, but, for most of the night, sleepers keep their spines straight using an extremely old technology: the pillow.

In ancient Mesopotamia and Egypt, pillows were rigid and were made of carved wood or stone. In ancient China, pillows were more typically ceramic. The soft, stuffed version of pillows most familiar to us was first recorded in ancient Greece, and they have diversified enormously over the past two and a half millennia. One contemporary website offers more than forty types of bed pillows that vary in size, shape, and material. We all share the common human dilemma of how to rest the head horizontally at night, but our specific pillow preferences are deeply individual.

In the morning we resurrect our bodies and prop our heads comfortably and almost effortlessly on our vertical column. We stand there, maybe for a second or two, and then we start a series of head movements that are almost continuous for the rest of our waking hours. Our neck not only gives us posture and pose, it gives us panorama and gesture. The cervical spine supports the head in three dimensions; the surrounding muscles of the neck give the head its fourth dimension: movement through time.

CHAPTER THREE

Panorama & Gesture

Moving the Head

To see the world, we move our heads a lot, about every six seconds. The specialized musculoskeletal anatomy of the neck makes it one of the most mobile and flexible regions of the body. In humans as well as most necked animals, this mobility plays a crucial role in expanding our visual range, our panorama, and our daily sense of what is happening all around us. Indeed, this expansion of the sensory world was likely one of the earliest and strongest forces shaping the evolution of the neck. But as so often occurs during evolution, new functions can evolve on top of old functions. Repeatedly, over evolutionary time, the neck was co-opted as a tool of communication and gesture. Many creatures and virtually all humans have a lot to say with their necks, though species and cultures commonly evolve their own language of head movements. In much of Europe and the United States, for example, people nod their heads up and down to say yes, rotate them left and right to say no, bow their heads to pray, and tilt their heads to kiss, but in other cultures, people have entirely different languages at the neck. Moreover, many animals use elaborate ritualized head movements and neck displays in courtship and aggression. Sometimes the neck is more about broadcasting your internal world than perceiving the external world.

Try sometime, for a few hours or even a few minutes, to live the life of a person with fusion of the cervical spine, the life of someone who does not twist or bend their neck. You will quickly notice that sitting still, you can only see a small slice of the world. To expand your horizon, you must move your torso or even your whole body. You might also find you elicit quizzical stares from people, looks that are more perplexed than if you were immobile at any other joint.

Rex was a rancher who owned a big expanse of land in the high desert of Central Oregon, where he allowed several of us biologists to conduct field studies. In our first encounter I was walking at the side of a dusty, lonely back road on his ranch. As his pickup truck came barreling down the road, I noticed it was fitted with an unusually large and expansive set of rear-view mirrors. Rex slowed to a stop, rolled down his window, and greeted me with his slow, folksy drawl, all while looking straight ahead through the windshield. Occasionally he glanced at me out of the corner of his eyes. As our chat concluded, he rotated his whole torso toward the window and rocked it forward slightly as a farewell gesture. His head stayed straight forward on his shoulders the whole time. Despite Rex's friendly words, I could not help but react to his unusual, rigid movements. Later, I learned he had undergone surgery to fuse his cervical spine following an injury. His elaborate system of mirrors expanded his view to allow him to see a full panorama from his driver's seat. But not even his generous cowboy charm could overshadow his immobile, expressionless neck.

PANORAMA: MOTORS AND SWIVELS

All our panoramic views of the world as well as our expressive head movements are enacted through a complex system of bones and muscles. The neck (cervical) region is by far the most flexible portion of the spinal column. Such flexibility is possible because of the large number of mobile elements and surfaces in the neck. In a span of about 30 centimeters (12 inches), seven vertebrae are joined at a total of thirty-seven surfaces, making the cervical spine the most complicated articular system in the body. These joints are controlled by more than twenty sets of muscles. Although the head is balanced mostly symmetrically on the spine, the neck muscles are far from symmetric. Muscles that extend the head backward are larger and more numerous than those that pull in other directions.

This concentration of muscles on the back of the neck is an evolutionary holdover from our quadrupedal ancestors who had to exert considerable muscular force to hold up and raise their heads. Some muscles of the neck are short and pass from the skull to the first two cervical vertebrae or between adjacent vertebrae.[1] Other neck muscles are longer and span many more cervical vertebrae or even insert onto vertebrae of the torso or beyond. In general, long muscles accomplish large movements of the neck while short muscles stabilize the neck and determine which specific joints can bend when the long muscles exert their pull. Long muscles overlay the short ones. Most of the shorter, deeper muscles arise early in the embryo and have changed relatively little over vertebrate evolution, while most of the longer, more superficial muscles originate later in the embryo and, evolutionarily, are highly derived modifications of muscles that once controlled the movement of gills in our fish ancestors.

One of the most interesting of these long superficial muscles is also one of the most visible muscles in the neck. Look in the mirror and turn your chin all the way to the right. There, on the left side of your neck is a bulging strap muscle, the sternocleidomastoid. Its long name gives away its attachment sites on bones. Originating on the breastbone (sterno-), it is a paired muscle on either side of the neck that forms the edges of the V-shaped notch at the base of the throat. It also attaches to nearby portions of the collarbone (cleido-). It spirals up diagonally on each side of the neck to insert on the bony knob (mastoid) just behind the ear. If you contract only one side of the sternocleidomastoid while relaxing other neck muscles, your ear moves toward the breastbone and your head will rock to the side and rotate a bit. If you contract only one sternocleidomastoid while stabilizing the head with other muscles,

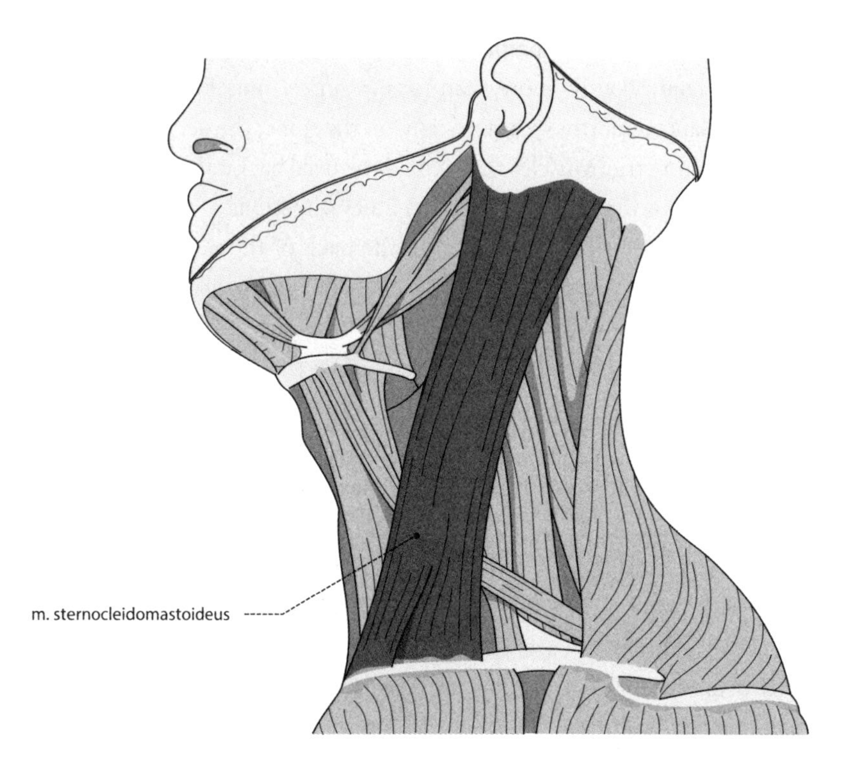

Figure 6. Musculature of the neck. Arrow points to the sternocleidomastoid muscle. Illustration by Olek Remesz, 2007, modified from Henry Vandyke Carter, ca. 1858 (Gray 1988).

your head will simply rotate horizontally toward the opposite direction. (This is what you did looking in the mirror.) If you simultaneously contract both sternocleidomastoids, you lower the chin downward toward your chest. This one muscle, activated in different ways, can move the head in three directions.

The terms that anatomists use to describe head movements are the same ones that aeronautical engineers use to describe the motions of an airplane. The head, like an airplane, moves with three degrees of freedom: pitch, roll, and yaw. Pitch is movement up and down (nodding "yes"), roll is rotation (gesturing "no"), and yaw is tilting left and right. Like the shoulder and hip joints, the neck allows motion in almost all directions. However, the broad range of movement is accomplished very differently at the neck. In the shoulder and hip, all movements occur at a single ball-in-socket joint. In the neck, different movements occur at different joints. The left/right tilt, or yaw, occurs through small movements at each of the five lowest joints down the cervical spine. The other two directions, pitch and roll, occur at two separate joints in the cervical spine. Pitch occurs at the joint between the skull and the first cervical vertebra (atlas), while roll occurs at the joint between the first two cervical vertebrae (atlas and axis).[2]

In the early nineteenth century, the functional elegance of these cervical joints took on theological significance, featuring prominently in the "Intelligent Design" argument for the existence of God. William Paley (1743–1805), often considered the father of Intelligent Design, saw evidence for God manifest throughout the order and function of Nature, but nowhere more plainly than in the ingenious anatomy of the cervical vertebrae. Paley likened the swivels of our "God-given" neck to the mechanism that permits the three-dimensional motion of a mounted telescope. In looking at the similarities between these anatomical and mechanical apparatuses, no one could "doubt the existence of counsel and design," he maintained.[3]

PANORAMA: AWARENESS AND STABILITY

It is probably not coincidental that Paley compared the mobility of the human neck to the swivel of a visual instrument, the telescope. Indeed, a discussion of neck function must include—perhaps even begin with—a description of our visual system. For humans, much of our neck movement serves our visual needs, except not for telescopy but rather for depth perception. Placing the eyes close together on the head allows for broad overlap in the visual fields of each eye. The brain then uses the slight difference in each view to compute the distance of objects in the world: the greater the difference between the views, the nearer the object. With two forward-facing eyes, we have the great advantage of depth perception, but in any one snapshot we can detect only a small slice of the whole world. To get a better sense of this restricted range, imagine conversing with your friend at a typical distance of 1.2 meters (4 feet). With eyes fixed straight ahead, you can see a highly resolved three-dimensional image of your friend's face that is about 30 centimeters (1 foot) in diameter, surrounded by a low-resolution border that extends an additional 15 centimeters (6 inches). The rest of the world is either very blurry or completely invisible. Put simply, with stationary eyes and no swivels, you can only see 40 percent of the visual world, and a large portion of even that is fuzzy.

Fortunately, our eyes can move within our head, and our head can move on our shoulders. Working together, these swivels enable us to view about three-quarters of the horizon. We constantly move our eyes and head to survey the environment, and so we see the world more as movies shot from a roving camera than as snapshots from a fixed tripod. The way in which our visual system is expanded by our musculoskeletal system is a clear example of how our senses always work in the context of movement: the breadth of our vision is possible only because we instruct our body to move. Our muscles and bones—especially those of our neck—give us our panorama as much as our eyes.

Although our swiveling necks and eyes broaden our visual world, they also greatly complicate a different visual task: how to re-create a stable world in our mind's eye. Gyrating our head and eyes in three dimensions could drive us nuts, like watching a video clip of an amateur videographer tracking an active child or pet. Swiveling systems of orientation can be really confusing and almost impossible to watch. Somehow, we have to stabilize the image.

• • •

From his videos posted on YouTube, Destin looks like a backyard tinkerer. Many people tinker with wrenches and screwdrivers; Destin tinkers with miniature video cameras and chickens. From his home in Alabama he shared with the world a video of his chickens executing a remarkable neck behavior: extreme head stabilization.[4] Many animals use their neck muscles to stabilize their head when their body moves, but few do it as well as chickens. In this video Destin holds a chicken around its middle and rotates its body in every direction, fore and aft, back and forth, up and down. The chicken's head stays in exactly one place the whole time. Perfect compensation. At the end of the video Destin adds an apparently throwaway line: "seems like it would be a good steady cam."

Two years later, Destin tried out this idea of a "chicken cam." Filming from his kitchen, he dons a pair of chemist's goggles to protect his eyes from the rooster's claws and straps a matchbook-sized video camera on the comb of the rooster. The main camera filming the video is focused on Destin and his rooster while a small inset video shows the world from the perspective of the rooster himself. Again, Destin rocks the rooster around in every direction, but the "chicken cam" streams a remarkably steady picture of the world. Without explanation, the video closes with "Great are the works of the Lord studied by all who delight in them. Psalms 111:2." To many people, chicken necks, at least those stuffed into the body cavity of a roasting bird, are somewhere between disgusting and odd. In Destin's kitchen they are nothing less than one of the Lord's great works.

• • •

We humans, like chickens, carry with us both the blessings and the burdens of a highly mobile mind's eye. Walk into most any technology convention and the remarkable virtual reality demonstrations will convince you that we delight in feeding our brains with panoramic images in three dimensions. Stumble into bed after too much alcohol and the spinning ceiling will convince you how unpleasant and disorienting it can be when the fixed world is unstable. Fortunately, humans have highly reactive systems that compensate and stabilize the world. Two sensory systems—neither of which are even included in the proverbial "five senses"—are recruited to accomplish this compensation. They are the vestibular and proprioceptive systems.

The vestibular system, located in the middle ear, detects the position of the head relative to gravity and the direction and acceleration of head motion—"which way is up?" and "how am I moving?" While the vestibular system supplies information about the head in space, it alone tells us nothing about the position of the head relative to the body. To have a sense of our bodies in the world and to re-create a stable world in our mind despite the swiveling of our eyes and head, we constantly merge information from our eyes and vestibular system with information from the proprioceptive system. This sensory system has receptor cells in muscles and joints all over the body. These receptor cells convey information about how much a muscle is stretching or contracting and how much a joint is bent. They give us our "sense of self" ("proprio" = self). While neck muscles serve as motors to swivel the head in every direction, they also have sensors to send information back to the brain to tell it about the position and motion of the head. Muscles without such sensors would be as limited as sensors, such as eyes, without muscles.

To illustrate how visual perception depends on eye-neck coordination and proprioceptive feedback, consider the details of this extremely mundane event. A vague flicker in "the corner of your eye" catches your attention, and you have the impulse to investigate. After detecting the motion on the periphery of the retina, the brain instructs the eye muscles to move the eyes so that the image will fall on the fovea, the most sensitive and centrally located part of the retina. The brain directs the neck muscles to start rotating your head toward the flicker while your neck proprioceptors provide feedback to the brain to guide the head in the correct arc and to slow it down as you approach the target. Soon you face the flicker, the image stabilizes on the fovea, and you can easily resolve the object: it's that fly again. All this in less than a second to accomplish a ridiculously mundane operation.

With almost every head motion the eyes move in a coordinated and compensatory way to keep the visual world stabilized on the high-resolution part of the eye. These adjustments are continual, fast, precise, and crucial, and they occur without a thought or conscious calculation. Otherwise, we would live in the world of a jerky video and be cognitively exhausted. The importance of this proprioceptive, visual, and vestibular coordination is borne out in neuroanatomy: the greatest density of proprioceptors in the body is in the neck, and the most densely innervated muscles in the body are the muscles of the eye socket. Keeping in mind all these processes underlying the mundane orientation toward a fly, now imagine a goalkeeper lunging for a soccer ball or a cat chasing a darting mouse.

Evolution has endowed us with the capacity to compensate second-by-second for the fact that our eyes—the movie cameras of our lives—are mounted in two sets of swivels. As organs of movement, our necks flex in all directions to expand the breadth of our panorama; as organs of proprioception, our necks root us in the fixed world. As they do so, we can appreciate our necks for both our most sublime experiences (e.g., scanning majestic vistas from the mountaintop) and our most mundane abilities (e.g., looking for our car keys).

• • •

The integration of proprioceptive, visual, and vestibular information—accomplished effortlessly and without a thought—is demonstrated in stories of two people who overcame malfunction in these systems with iron-willed intention. In 1960 the physician John Crawford recounted his own experience of the damage to his sense of balance after treating himself for a bout of tuberculosis.[5] The antibiotic he took, streptomycin, accidentally poisoned crucial cells in the organs of balance in the inner ear. At first, even reading was impossible. Any movement of the head while he was studying his medical texts caused the words to jiggle in his field of view. Crawford compensated by immobilizing his head between two metal bars on the frame of his bed. Even after he partially recovered, his persistent vestibular deficits forced him to adjust his social habits as well. He adopted a practice of saying hello to everyone he passed on the street. The simple act of walking jostled his head, and without a vestibular system informing his neck and eye muscles how to automatically adjust for these bumps, he could not stabilize his vision well enough to perceive individual faces. He could not tell which passersby were familiar and which were strangers, so he greeted them all. Crawford never fully regained his vestibular sense, but he learned how to navigate the world using other sensory information, eventually even relearning such complex activities as tennis and swimming.

Ian Waterman was blessed with at least two indispensable traits: an indomitable pride not to be defeated by his rare neurological disease and a healthy set of neck proprioceptors. In *Pride and a Daily Marathon*, Jonathan Cole tells the story of Waterman, who, at age nineteen, taught himself how to move after a viral infection disabled much of his peripheral nervous system.[6] The nerves that activated Waterman's muscles were not damaged. However, the nerves carrying sensory information from his muscles and skin were selectively destroyed, and without sensory feedback from his muscles, Waterman had no idea how to initiate or execute movement. At first, his attempts to move caused erratic and uncontrolled motions. When reaching for a cup, his arm might fly up above his head or even accidentally swing to hit his companion. Without most of his proprioceptors, he was effectively immobile. Waterman painstakingly regained mobility by substituting vision and other senses for proprioception. He began all his movement with conscious calculation and continual trial-and-error visual feedback. While executing any motion, he watched himself to make sure each body part actually moved as planned and then corrected any "disobedient" motions. Even the simplest of actions took almost all his attention and mental effort.

Through years of determination and practice, Waterman achieved a remarkable range of mobility, including walking. Nevertheless, without the ability to sense his body's position in space automatically and unconsciously, his daily physical activities were marathons—not of muscle fatigue but of mental exhaustion. Waterman's efforts were crucial in regaining mobility, but his recovery would likely have been impossible if the disease had not spared the proprioceptive nerves above his shoulder. So, while he learned to consciously execute desired movements by watching where his limbs moved relative to his torso, his functioning neck proprioceptors allowed him to have subconscious awareness of where his torso was positioned relative to his eyes and to gravity. This sense of orientation provided by the neck was one set of exceedingly complex calculations Waterman did not have to make. Neither vision nor movement would be possible—or at least easier than a marathon—without two senses that are so automatic that they barely enter our awareness. With our eyes and (most of) our muscles segregated into different parts of the body, the neck that connects them and its abundant proprioceptors play an essential but seldom recognized role.

PANORAMA: VISION AND AWARENESS

Because the neck integrates our senses of vision and balance located in the head with the rest of the body, it helps give us our sense of self and place in the world. But crucially the neck also affords us the capacity to know about the world beyond ourselves, allowing us to see from different angles and perspectives, to view the world below, above, and on either side. Combined with eye and torso movement, we can see panoramically. With a rigid neck, our knowledge of the world would be limited indeed.

For Plato, an immobile neck was nothing short of an allegory for the inborn limitations of human knowledge. He likens humans in their native state to prisoners in a cave who, since childhood, have had their "necks chained so that they cannot move, and can only see before them, being prevented by the chains from turning round their heads."[7] In front of them, the prisoners can see moving images on the wall of the cave. But with their heads restrained, they cannot see the source of these images. A fire burns behind the prisoners and, between the prisoners and the fire, puppeteers raise objects whose shadows are cast ahead on the wall of the cave. The puppeteers are hidden, so they cast no shadow. Confined to tunnel vision by their shackled necks, the prisoners cannot turn to see the cause of the shadows behind them, and they mistake these shadows for the real world. Their immobile necks make them a captive audience to a grand illusion. We can only know the reality behind our perceptions by twisting our heads.

• • •

Owls are no doubt one of the most perceptive of all creatures. They usually perch high in trees and scan their environment by twisting their remarkably flexible necks. With barely any commotion, they gaze all around themselves and perceive the whole world. Surely such silent omniscience has contributed to their reputation for wisdom. In the West, owls have been considered wise at least since ancient Greece, when Athena, Goddess of Wisdom, adopted them as her favorite. In English folklore owls were cloaked in mystery and darkness. According to one legend, an owl will eerily follow you with its eyes even to the point of suicide; it will wring its own neck if you walk around it in circles.

To an anatomist, there is little mystery about why an owl needs such a flexible neck. It is the natural consequence of a constrained ability to swivel at other joints in the body. They are so flexible at their necks because they are so rigid in the rest of their bodies. To fly, all birds, including owls, have rigid trunks that twist very little. However, birds vary widely in their neck mobility and visual breadth. As one example, woodcocks are short-necked birds that forage in the soil for earthworms and other invertebrates. Vision plays little role in locating prey, but their vision is well suited for detecting predators. With eyes on the sides of their heads, woodcocks can see 360 degrees when holding up their heads. Even with a stationary head, they can watch all around to spot predators that may be watching them. But with only 10 degrees of binocular overlap, their depth perception is poor, so they themselves would be horrible visual predators.[8]

By contrast, highly specialized avian predators (such as owls and other raptors) have highly mobile heads with eyes directed forward. To hunt, owls must judge prey distance very accurately, so their eyes are placed frontally to maximize binocular overlap. Their visual field has 70 degrees of binocularity. Moreover, because they often hunt at night, their eyes must be large to capture as much light as possible. Since their eyes are both large and frontal yet still must fit into a relatively small head, their eyeballs are not balls at all. Imagine trying to fit two golf ball–sized eyes into one side of a softball-sized head, while still leaving room for a brain and a face. It just would not all fit. Instead, owls' eyes are cylinders that point straight forward. With this shape they cannot swivel in the orbits the way eyes of most vertebrates can. Owls' eyes are fixed and can only stare straight ahead. With a rigid trunk and immobile, frontal eyes, owls could see only a small slice of the world were it not for their remarkable neck that can rotate in a three-quarters' (270-degree) arc. Combined with a 110-degree visual angle, owls can see more than a full circle, without even shifting in their perch.[9]

It is clear why owls need fantastically flexible necks to lead their life as flying, nocturnal predators. However, a greater mystery has been *how* owls manage to twist their own necks so far and how they do so without wringing (overtorquing) their spinal cord. The flexibility of owls' necks is due in part to the shape and number of vertebrae. Owls have fourteen cervical vertebrae, twice as many as humans and other mammals but typical of other birds. The joints at the core of the vertebrae (the centra) are saddle-shaped and can rock easily in all directions. The connections at the more lateral projections (the zygapophyses) are relatively short and loose. Perhaps more important, the vertebrae vary in orientation from head to torso, giving an overall curvature to their spine. This curvature reduces the amount of twisting that the spinal cord must withstand.

When an owl scans its environment, it appears to swivel its head in a single smooth movement. However, the extreme rotation of its head is actually accomplished by distinct regions of the neck moving in different orientations, rather than uniform motion along all the vertebrae. The owl's neck is not a twisting column. Instead, its cervical spine is

S-shaped, with the upper and lower portions moving in different directions and through different arcs. To picture the complex movement of the owl neck, try this simple exercise. Hold your arm straight in front of you parallel to the ground, then bend your forearm vertically by flexing your elbow into an L-shape and, finally, fold your hand horizontally at the wrist. In this position your hand is the owl's head, your forearm is the upper portion of the owl's neck, and your arm/shoulder is the lower portion of the owl's neck. Now, twist at the elbow, and notice that your hand—"the head"—rotates almost half a turn (a 180-degree arc). Then, swing your arm side-to-side, from midline outward (laterally) at the shoulder. You can see that this also moves your hand in a rotational arc, about a quarter turn (90 degrees). If you combine the two motions, "the head" can rotate through a much larger three-quarters' arc (270 degrees).

The analogy of this arm movement to the motion of an owl's neck is imperfect, but it nonetheless illustrates how a large rotation can be accomplished by combining two forms of movement occurring at two locations. In owls the head bends forward from the spine to form the upper curve of their S-shaped cervical spine, and near the thorax the spine bends inward horizontally toward the body, creating the lower curve of the S. By far, most neck rotation in owls occurs near the head between the first four vertebrae, where the motion is mostly twisting (yaw) and a little lateral bending (roll). By contrast, the lowest and most horizontally-oriented part of the neck does very little twisting. After all, twisting between these horizontal vertebrae would cause the head to keel over to the side. Instead, these lower vertebrae contribute to head rotation by swinging the base of the neck laterally, away from the midline. (Recall the swinging motion of your arm.) As a consequence of this side-to-side motion, the head does not rotate on a single twisting axis directly over the body. Rather, the owl's S-shaped cervical spine displaces the head in front of the body axis. By adding this lateral bending of the vertebrae at the base of the cervical spine to the twisting motion above, the spinal cord within the vertebrae is spared the full 270-degree torque.[10]

The extreme flexibility of the owl's neck presents an additional problem: how to rotate so far without rupturing the crucial blood vessels in the neck. Humans can turn their heads about 90 degrees, but we face a potentially lethal geometry if we turn it any further. Because the major arteries to the brain (the carotid and vertebral) are several centimeters away from the axis of rotation, they become stretched as the neck rotates. The walls of these arteries are fragile, and in humans they can rupture with too much twisting. How then does an owl turn its neck so far without damaging these vital arteries? The best clues came from a research team led not by an ornithologist or an anatomist but rather by a medical illustrator.[11] As a graduate student, Fabian de Kok-Mercado wanted to illustrate the circulatory system of an owl's neck. He injected dye into the cervical arteries of owls while turning their heads in an X-ray scanner. As in other backboned animals, these arteries pass through holes in the cervical vertebrae. However, in owls these holes are especially large—ten times the diameter of the artery—with air spaces surrounding the delicate arteries. So, as the owl's arteries ascend through the neck, they have more room to stretch and more cushioning during head rotation. In addition, the two major arteries to the brain have several unusual collateral arteries that can reroute blood flow if either is pinched by neck torsion. Underlying an owl's mysterious twisting neck is an adaptive design that seems to bear witness to a certain functional wisdom.

• • •

With all the apparent advantages of a long flexible neck to scan the environment, it is curious that some animals have especially squat necks or necks that greatly limit head mobility. Some species go so far as to fuse cervical vertebrae. In such "syncervical" necks, usually two or three but as many as seven vertebrae are fused together. Such fusions have evolved independently at least twenty times among vertebrates, including in mammals (many marine mammals, armadillos, pocket gophers, kangaroo rats, and porcupines), birds (the hornbill), and extinct reptiles (many aquatic and terrestrial dinosaurs).[12] These diverse groups have drastically different bodies and behaviors, but they share the similar need to stabilize their heads against unusually large forces.

Cervical fusions are most commonly found in burrowing mammals that use their heads as shovels. Moles, armadillos, and pocket gophers dig tunnels by lifting their heads with great force using muscles on the back of their necks. One champion digger, the mole rat, can lift fifteen to twenty times its body weight by raising its head. Researchers have hypothesized that fusion of certain cervical vertebrae effectively enables a portion of their necks to move as a single unit. Just as a long-handled shovel can pry with greater force, this fused neck can better lift the digging head. In contrast to these bulldozer animals, a second category of animals with fused necks are especially light and locomote by hopping. Kangaroo rats, for example, bounce along on two legs with their bodies tilted almost horizontally. They commonly stuff their cheek pouches with seeds, which gives their head added weight. A long neck extending the heavy head far forward would be difficult to balance over two hind legs, so their necks are exceptionally short. Moreover, during hopping, the pounding upward force from the legs combined with the downward force of gravity on the heavy head places a lot of stress on the cervical vertebrae. Fusing these vertebrae together likely helps prevent the bone from failing.

In addition to these small terrestrial hoppers and subterranean diggers, underwater behemoths, the whales, have relatively short or fused necks. In the case of whales and dolphins (cetaceans), neck reduction and cervical fusion help keep their heads stable while locomoting through water, which exerts a strong frictional force against the head. In addition cetaceans have little need to turn their heads while foraging because most of them are not primarily visual predators. Notably, the cetacean species without fused cervicals (narwhals, beluga whales, and certain river dolphins) are those that must navigate through waters filled with many obstacles (icebergs or trees) and thus may require more flexible necks to peer around corners.

Terrestrial giants like elephants and big-headed dinosaurs such as triceratops do not so much contend with frictional forces, but they must counteract enormous gravitational forces on their massive heads. Elephants have large jaws and teeth for chewing vegetation as well as heavy tusks, and they also use their trunks to pick up heavy loads. Triceratops had large bony shields that added to the mass of their heads. For both these giants, reducing the neck length and fusing some of the vertebrae enables them to hold up their heavy heads but limits their range of head mobility. Since both species have their eyes on the sides of their heads, there is less need to move their heads for visual breadth. At more than 5,000 kilograms (13,000 pounds), it is not so crucial to scan the world for predators.

Overall, highly mobile necks have proven very useful in the evolution of terrestrial vertebrates, but the hoppers and diggers as well as the behemoths of land and sea illustrate that evolution can have "second thoughts" about the neck. For these creatures the necessity to resist friction and gravity overrides the advantages of panoramic awareness from a broad visual range. In shaping these animals, evolution did not simply revert to the necklessness of fish. Instead, it secondarily applied its own form of cervical fusion.

GESTURE: HEAD MOVEMENT IN COMMUNICATION

Humans in all cultures use their necks to communicate nonverbally, but the meanings of such movements do not always translate smoothly between cultures. I remember such confusion during one of my first transactions abroad, while living in Greece. At the bus station I mustered up whatever courage and rudimentary Greek I had and nervously asked the clerk, "Does the bus leave this morning?" He raised his eyebrows and gave a quick upward tilt to his head. No words. Did this half nod mean "yes"? Utterly stumped, I decided just to wait and see. After a few hours of waiting, I discovered that this nod evidently meant "no." Head twists can also be misinterpreted across cultures. Many years ago, two of India's most popular singers, the Mishra brothers, Rajan and Sajan, performed in a German variety show. One brother sang masterfully while the other brother looked on, shaking his head from side to side, a gesture of deep admiration common in India. Afterward, the show's host complimented the singer for his masterful performance and asked the other brother, "While your brother was singing, why were you disagreeing?"

Beyond communicating—or miscommunicating—specific information, head movements also humanize people and augment conversation. Listeners often rock their heads to convey attentiveness; speakers gesture with their heads to punctuate and modulate their words. One of the first lessons offered to television newscasters—"talking heads"—is to move their heads subtly but regularly.[13] Newscasters with stationary heads look "stiff, scared and nervous." Watch a television news program with mute on, and it will become evident that successful television personalities communicate with visible motions of the head as well as audible motions of the vocal tract.

When people are unable to move their neck because of surgical fusion, it significantly limits their expression and impacts their social interactions. However, the opposite extreme—erratic uncontrolled movements of the neck—can be more disruptive to the afflicted and confusing to others. Some people with a movement disorder termed cervical dystonia spend their days and nights with uncontrollable tremors in the neck muscles. They may experience an unwanted twist or tilt of the head every few seconds or their head may even lock into position at a tilted angle for hours or days at a time. Often they cannot read, keep their balance, sleep well, or maintain work. Sometimes the pain is excruciating, "an ice pick in the neck," reported one patient. It is hard to imagine the torment. In 1907 the Romanian artist Constantin Brancusi, who was known to be intrigued by human malformations, sculpted a boy with an extreme lateral flexion, torsion to his neck, and a drastically elevated shoulder—a posture common in cervical dystonia. The work is simply titled *Suffering*.

With cervical dystonia, all your intentions to scan your world and express yourself are superseded by an uncontrollable set of internal signals. To lose control of one's head is to lose control of oneself, and with that, to lose fluid social communication. For some people with cervical dystonia it is an enduring task to avoid miscommunicating with their heads. Their body language seems erratic and confusing to others. "I receive a lot of attention from people trying to understand what they're looking at," writes Cheryl Dillon, who has suffered from cervical dystonia for years. "'Are you on the phone?' 'Are you having a seizure?' 'Are you really going to drive?' While the public's trying to understand what they're seeing, I'm trying to live inconspicuously."[14]

The causes of cervical dystonia are not clear. Occasionally, it arises following injury or stroke, but more commonly it emerges without an incident. The condition appears to have a complex genetic component and likely involves hyperactivity or miscommunication between brain regions that control movement. To quell the spasms in the short term, patients often discover "sensory tricks"—for example, light touching on the cheek, pressing on the head, clenching hands together, or visually fixating on the head while looking in the mirror. Such "tricks" can transiently ease the spasms, but for longer-term relief, doctors often turn to injections of botulinum toxin to temporarily reduce spasms in the inappropriately contracting muscles. Such injections generally alleviate symptoms for three to six months. In cases that do not respond well to pharmacological treatments, patients may receive a more invasive procedure termed "deep brain stimulation" in which electrodes are implanted in movement-associated brain regions to deliver chronic, individually-tuned electrical stimuli.

While botulinum toxin treatment and deep brain stimulation have helped many patients, some have turned to movement therapies. Federico Bitti, who told his story on a widely circulated TED Talk, sought movement therapy with Joaquin Farias, the director of the Neuroplastic Training Institute in Toronto, and their work together got Bitti started down the path to recovery.[15] But it was the music of the pop singer

Madonna that first sparked his rediscovery of how to move within the confines of his dystonia. At age thirty-two, Bitti worked as a journalist in Italy. One day during an interview, he felt his neck uncontrollably deflecting away from the person he was interviewing. Embarrassed and scared, Bitti quickly learned that he could use his hand to turn and hold his head in the right direction. But after the interview the problems grew worse. His initial stiffness in the neck turned into a tremor. The only way he could control the shaking was by bracing his head against his elevated shoulder—the posture of Brancusi's *Suffering.* The dystonia eventually spread to his back and severely compromised his walking.

For six weeks Bitti and Farias worked together doing movement exercises, focusing mostly on strengthening the weakened muscles that counteracted the hyperactive muscles. "In those days, I learned, or rather relearned how to be in charge of my body again." Through these therapeutic sessions his neck and his whole life began to turn around. However, the real "revolutionary change" came on his own. Once while walking down the sidewalk listening to Madonna's "Vogue" through earbuds, Bitti noticed that his gait greatly improved. Walking turned into dancing, and he was astonished to feel that when he moved rhythmically to the beat, the symptoms of his dystonia almost disappeared. When Farias saw this, he exclaimed, "Oh my God, this is your treatment." Regular dancing became part of Bitti's life. "If dystonia was my hell," he said, "[dancing] was my paradise." Dance not only helped Bitti regain control of his neck, but it also allowed him to take back his life and experience the ecstatic joy of expressive movement.

Beyond Bitti's individual experience, published research has confirmed that various forms of movement therapy, particularly in combination with botulinum toxin injections, can help ease the tremors and tribulations of dystonia. Patients with cervical dystonia have commonly reported a release from dystonia symptoms when dancing. In therapy, dance can help people learn and relearn how to move their heads and thereby their whole bodies. On the dance floor or stage, dance reveals the expressive potency of head and neck movements.

GESTURE: HEAD MOVEMENT IN DANCE

At any pop music concert you will find many in the audience dancing in their seats as they bob their heads to the beat, something we do almost instinctively. Human infants and even some birds do it. But head motions in more formal dance are often intentional and choreographed. Formal dance styles from all over the world incorporate some sort of head motion, although they differ notably in how dancers move the head, both in conjunction with whole body motions and as a separate mobile element.

In classical ballet, head motions are generally highly controlled and aligned with the motions of the rest of the body. Rather than acting as a source for expressive motions of the head, the neck serves mostly to highlight ballet's ideals of elongation, balance, and smooth movement. Students of ballet are often told to imagine a string attached to the top of the head pulling it straight upward. The head and neck should align directly above the torso. At the same time, the shoulders are pulled down into a "dripping" position, the collarbone is thrust forward, and the overall effect is to further emphasize the ideal of a long neck extending from a long body. The head rarely tilts drastically in any direction.[16] To do so would compromise a sense of balance. When the head turns at all, it is usually in conjunction with other body parts so as to join—or at least not break—the "grand linear curves." The head may turn to track the movement of the arms or tilt forward or backward to counterbalance movements of the legs. Rarely does the head initiate movements or move independently. Different schools of ballet have their own variations, but in almost all forms the neck moves subtly, gradually, and in alignment with the body to create long smooth lines.[17]

Against the tight constraints of ballet, modern dance rose like a phoenix in the early twentieth century. Choreographers such as Isadora Duncan, featured in the preface, as well as Loie Fuller and Martha Graham were determined to showcase movements that were foremost expressive and individual, rather than idealized and controlled.

While ballet sought smooth curves and symmetry, modern dances often emphasized angularity and asymmetry. In Graham's 1914 piece *Spectre*, the dancer often separates the movement of the head and the torso. She lunges her chest forward, but her head lags behind; she starts a full-body rotation by first twisting at the neck and then her torso follows; she pulses her head rapidly back and forth as she slowly elevates her torso up from a backwardly inclined position. It hardly looks effortless or fluid. Many of Graham's dances have a "whiplash intensity in the belief that life itself is effort . . . [filled with] agony and rapturous exaltation," writes dance historian Jack Anderson.[18] The neck contributes to the expressive fervor by partially disrupting the line of the body, enabling the head its own movement as an expressive element.

• • •

Centuries before the revolutions of modern dance, classical Indian cultures developed dance forms that equally relied on a highly expressive neck. Their importance was codified in a Sanskrit treatise, the *Natya Sastra* written in about 100 BCE, that formalized the structure and philosophy of all performing arts.[19] The *Natya Sastra* delineated four categories of neck movements (*greeva bheda*) and nine categories of head movements (*shiro bheda*). In the Sundari *greeva bheda*, for example, the neck moves laterally, and the head slides left and right while facing forward. This movement commonly signals the beginning of a friendship or expresses admiration. In the Paravrittam *shiro bheda*, the head rotates in a quick turning motion. As an abrupt snap, this gesture is commonly used to grab the viewer's attention or to punctuate the conclusion of a series of motions.[20]

One central aesthetic component of classical Indian dance throughout its history is the concept of *rasa*.[21] Translated literally, it means juice or nectar or extract. Rasa is the essence or emotional flavor of a performance, and dancers strive to infuse the audience with this feeling. The rasas are systematically categorized in nine ways: erotic, humorous, pathetic, terrible, heroic, fearful, odious, wondrous, and peaceful.

Figure 7. Kathak dancer. Photograph by Parwati Dutta, 2009.

Each of these rasas has its own emotion, deity, color, and element as well as its own characteristic head/neck movements. To express the erotic rasa, which includes romantic love in general, the dancers use the Sundari head slide. For the humorous rasa, the dancers may bobble their heads in all directions to convey the silliness of the situation. The neck is so fundamental in the expressive vocabulary of Indian dance that one dancer wrote: "Like salt is to food, neck movements are to dance. Like fire is to the sun, neck movements are to dance."[22] As in ballet, there are many guidelines for the use of the neck in Indian dance, but rather than constraining motions, these codes infuse emotional flavor and color into the performance.

In addition to displaying this highly structured emotional vocabulary, the neck in Indian dances serves to guide the viewer's attention. Rachna Ramya, a scholar, performer, and teacher of Kathak dance, explains that the principal role of the neck is to convey direction in movement.[23] The dancer tilts her head down and to the side, stares down her arm, drawing attention to an important hand gesture. She stretches her neck up, turning her head in all directions to indicate that she is searching for her lover. Many of her motions across the stage are initiated by a quick turn of the head to point the way.

• • •

The dancer strides and skips across the stage, pointing her path with a twist of the neck just before she moves or changes direction. From your seat near the stage, you track the movements of the dancer by rotating your head. Your companion is especially impressed and tilts and turns her head toward you, with a telling facial expression. You nod in agreement. At the end of the piece the house lights come on. You scan the auditorium to appreciate its ornate architecture and, finally, to search for the nearest exit door.

We scan the world to see and gesture to the world to communicate. For humans our knowledge of the world relies on our capacity to see panoramically. The neck is a pivotal empirical tool. For many animals (and for humans during most of our evolutionary history), life's moment-to-moment behavioral decisions depend on seeing the predators, prey, mates, food, and terrain around them, not just the stuff in their stationary visual field. The neck is thus a crucial survival tool. In social species, especially humans, the mobile head has become an organ of communication and gesture. The neck drives a silent but potent body language that augments our more celebrated modes of communication—the voice from the throat and gestures from the hands and face.

CHAPTER FOUR

Tubes & Transport

Conduits to the Torso

Our bodies, like the buildings we inhabit, are compartmentalized. Our houses have designated rooms for cooking, eating, eliminating, and sleeping, and our bodies have regions dedicated to sensing, ingesting, digesting, breathing, and reproducing. But these compartments are hardly independent; they are connected by an elaborate tangle of plumbing, ducts, and wires that distribute fluids, air, and electrical signals from one specialized region to another. In buildings, most of these connectors run in the walls and floors. In our bodies, most run through our necks.

Arteries and veins transport blood between the organ of circulation (the heart) and the hungry brain tissues of the head. The esophagus transports food from the site of ingestion (the mouth) to the site of digestion (the stomach). The trachea transports air between the site of inspiration (the nose/mouth) and the site of respiration (the lungs). The nerves transmit signals between the site of perception and cognition (the brain) and the sites of tactile sensation and motion (the skin and muscles). Lymphatic vessels drain watery lymph from the head and neck and return it to the bloodstream near the heart. The neck is an exceedingly busy transit corridor with diverse sets of cargo, all running in parallel. The nerves and lymphatics are discussed in chapters 6 and

10, respectively. Here I explore the passage of blood, food, and air through the neck.

To get a sense of exactly how much traffic moves through our neck, let's do the numbers, or at least approximate them. Blood pulses from our heart to our head every second or so, with an average flow rate of about 375 milliliters—half a wine bottle—per minute or about 540 liters (give or take, 140 gallons) per day.[1] We breathe every four to five seconds, with about half a liter flowing back and forth with each breath. So, every day about 11,000 liters (roughly 2,900 gallons) of air is inhaled through the trachea. We swallow about six hundred times per day, usually just saliva or water and occasionally a slurry of macerated food. The volume of each gulp varies considerably, but on average we transport about 5 liters (1.3 gallons) of fluid down our esophagus per day. As a rough calculation, all together about 11,500 liters (3,000 gallons) of material weighing about 600 kilograms (1,300 pounds) pass through these conduits every day. And this is just one direction; blood and air flow in both directions. So give yourself credit. Even on your laziest days, you move well over a ton of material through your neck.[2]

Although it is difficult to imagine the daily total of all this movement, we can nonetheless readily perceive each moment of the flow. Each pulse of blood is palpable to the touch, each breath is audible, and each swallow is both visible and audible. Like most background activities in the infrastructure of our lives, all this flow is normally outside our awareness, but it is not hard to detect if we give it just a bit of attention. Usually we only consider this flow when a malfunction—clogged pipes or broken ducts—demands our immediate response. Unlike the flow in our buildings, the flow in our neck tubes can never be disrupted or postponed for deferred maintenance. The brain is a particularly hungry organ, consuming about 20 percent of the body's energy but constituting only 2 percent of its mass. Because the brain has such high energy demands and must function continuously for survival, we must constantly supply it with oxygen and eliminate its carbon dioxide waste through blood flow. Furthermore, the brain is a picky eater. Unlike most organs, which can use fats, proteins (amino acids), or sugar (glucose) and store these fuels locally, the brain runs only on glucose, and it stores very little fuel. The brain imports all its sugar energy through the bloodstream.

Given that this transport is an absolute necessity and has been so throughout vertebrate history, one might think that evolution would have nearly perfected the design of the transport systems by now. But we all instinctively know that this is not even close to true. The pipes and tubes in the neck are thin, narrow, and near the surface. They are prone to getting clogged, compressed, or cut, and we live disconcertedly close to choking, suffocating, or blood loss from the neck. Why would millions of years of evolution have yielded such a fallible design? Some vulnerabilities in our plumbing are the flip side of the neck's multifunctional capacities. The neck must heroically transport massive amounts of vital fluids while also being flexible and, in many vertebrates, vocal. These simultaneous functions sometimes conflict. "If the human body were a building, our neck would arguably be the most poorly conceived room in the house, overflowing with functionally mismatched organs stuffed there to accommodate other design priorities," writes Matthew Rozsa.[3] Our neck anatomy is a compromise among competing demands, not a perfect solution.

A second factor underlying our imperfect neck is that during evolution the new functions incorporated into the neck have always been built on top of old plans. Our multifunctional neck is a product of jerry-rigged renovation rather than optimized innovation. When new structures—for example, the trachea and the lungs—get spliced into preexisting structures (the throat), they open up all sorts of possibilities, such as air-breathing and terrestrial existence. But these added structures serve short-term needs, not eventual, predestined evolutionary goals. On the whole, such design by improvisation works well enough, and life persists and evolves with grandeur despite carrying certain remnants of awkward splicing. The transit corridors of the neck allow all the advantages of compartmentalizing separate functions in the head and the torso, but we are fated to live with the inevitable trade-offs and vulnerabilities embedded in our improvised evolutionary history.

BLOOD AND VESSELS

To be flexible, the neck must be thin. But one trade-off of this constriction is that the transport tubes end up close to the surface and vulnerable to puncture and pressure. For blood, the flow is tangible with a light pressure to either side of the throat. Indeed, this is the first place clinicians touch to check if a person is alive. If you press lightly on the carotid arteries in the neck, you detect life.[4] If you press too hard or long, you cause collapse and black out. Hippocrates, who named these arteries in the fourth century BCE, clearly understood this relationship between carotid constriction and mental function: "carotid" derives from the Greek root *karotis*, meaning "to stupefy." Occasionally plaque accumulates on the inside walls of these arteries and reduces blood flow to the brain. In these cases symptoms can include slurring of speech, deficits in memory and vision, confusion, and dizziness. Worse yet, plaque can detach from the carotid artery wall and lodge in a brain artery, causing a stroke and possibly permanent brain damage.

If constricting carotid blood flow makes us "stupid"—to draw on Hippocrates's word—the converse may also be true. Increasing blood flow through the neck probably makes us smarter, at least over evolutionary time. Two lines of research, one using cutting-edge genomics and one using old-fashioned anatomy, indicate that evolutionary increases in carotid blood flow are associated with (and likely contributed to) evolutionary advances in human cognitive ability. When researchers sequenced the gorilla genome, they were able to identify genes that show particularly high rates of evolution in the hominine lineage.[5] The gene with the single strongest evidence for such accelerated evolution was *RNF213*, and it was already hypothesized to contribute to blood flow to the brain. Naturally occurring mutations of this gene in contemporary humans are associated with Moyamoya disease, a rare condition in which cerebral blood flood is restricted and risk of stroke is heightened. Although the precise action of the *RNF213* mutation in causing Moyamoya is not clear, patients with the disease have excessive production of cells in the lining of the cerebral arteries, particularly the internal carotid artery, and this overproduction narrows the arteries and decreases blood flow rates to the brain.[6] Reversing the logic, researchers postulated that evolutionary modifications to *RNF213* in the hominine lineage might have enabled an increase in cerebral blood flow that was necessary to provide oxygen and glucose to the expanding brain volume during human evolution.

More recently, new fossil evidence indicates that enhanced blood flow through the internal carotid arteries may have played a role in evolution of human intelligence beyond simply feeding a larger brain mass.[7] Carotid arteries themselves do not fossilize. However, the holes through which they pass into the skull do. A team of researchers headed by Roger Seymour measured these holes in skulls from twelve hominid lineages and estimated the blood flow rate from their diameters. As expected, species with larger brains had larger holes and likely had greater blood flow. But surprisingly, over hominine evolution, carotid blood flow increased to a greater extent than brain volume. Across the 4.4 million years separating modern humans and their earliest human ancestors, brain volume increased about fivefold while blood flow rates increased about ninefold. That is, the brain's appetite for blood grew faster than its size. Given that blood flow is a good indicator of neuronal activity, this disproportionate increase in blood flow suggests that the rise of human intelligence depended more on the "density of brain activity" than simply on the size of the brain. We can thank, at least in part, our plumbing for our smarts.

• • •

To pump blood up to our hungry brain, we must work against gravity. In our upright stance, the brain is about 0.4 meters (approximately 1.3 feet) above the heart, and to pump blood upward this distance, the heart generates a blood pressure of about 120/80 mm Hg. If blood pressure falls much below this, even briefly, our brains do not receive enough blood and oxygen, and we might faint. By contrast, brief increases in blood pressure are not dangerous. However, if blood pressure stays high (say, above 140/90) chronically for months or years, it causes all sorts of chronic cardiovascular problems, including excessive growth of the heart and damage to the arteries.

Now consider the blood pressure issues in a considerably taller creature. For giraffes, their head is about 2.5 meters (8 feet) above their heart, and to pump blood up their near-vertical neck to their brain requires enormous pressure (220/180 mm Hg)—almost double that for humans. This exceptionally high blood pressure—a natural hypertension—has prompted both zoologists and clinicians to ask two questions: (1) How can giraffes generate such high pressures from their heart? And (2) with such high blood pressures, how do giraffes avoid the cardiovascular pathologies associated with human hypertension? The answer to the first question is surprising. You might imagine that creating such high pressures would simply require a large heart. However, in terms of heart size, there is nothing unusual about giraffes compared to short-necked mammals—that is, giraffes have about the same heart size as other mammals of their body size. In fact, the interior chamber of the left ventricle, from which the blood is pumped to the whole body, is relatively small. But the muscle of their left ventricle, which generates the major pumping force, is unusually thick. Thus each heart contraction ejects only a small volume, yet because the thick muscle can produce a lot of force, the pressure generated is very high.

Interestingly, these two unusual features of the giraffe heart—thick ventricular walls and low ejection volumes—also characterize the hearts of humans with chronic hypertension and heart failure.[8] The main difference is that giraffes have relatively low pressures during the relaxation phase of the heart cycle (diastole), which allows them to avoid much of the cardiovascular damage associated with heart failure in humans. The precise cellular processes underlying this resistance to heart pathologies are not yet fully known, but it appears that giraffes, unlike hypertensive humans, do not develop cardiac fibrosis, a condition in which certain cells (fibroblasts) respond to tissue strain by depositing excessive fibrous proteins in the spaces between cardiac muscle cells. Such fibrosis stiffens the heart muscle and reduces its ability to fully relax between contractions, but somehow giraffes avoid it.

The idea that suppressed fiber production may protect giraffes from cardiac pathologies of hypertension is supported by recent genetic studies of giraffes and their relatives.[9] These studies also suggest that changes in genes that regulate fibroblasts may have been a key target in giraffe evolution. In 2021 researchers examined the complete genome of giraffes and compared it to other mammals, including the giraffe's closest short-necked relative, the okapi. One of "the most conspicuous targets of selection" in the giraffe was the gene *FGFR1*, which codes for fibroblast growth factor R1, a molecule important in regulating fibroblasts. Mutations in *FGFR1* in both humans and mice cause cardiac malformations, suggesting that it is normally involved in the formation of fibers in the heart.

But does the unique giraffe version of this gene confer any protection against the cardiac pathologies associated with hypertension? To address this question, the researchers created segments of DNA that had the same sequence as the giraffe *FGFR1* gene and, using CRISPR gene-editing technology, substituted it into the DNA of mice. Baseline blood pressure did not differ between mice that had the giraffe version of the gene and those with the mouse version, indicating that the gene does not likely contribute to the unusually high blood pressure found in giraffes. However, when the mice were chronically given a hormone that increases blood pressure, those with the mouse version of the gene showed several cardiac pathologies associated with hypertension, including fibrosis and low ejection volume, while those with the giraffe version were pathology-free. This was strong evidence that during giraffe evolution mutations to the *FGFR1* gene enabled their hearts to withstand the pressures that were required to pump blood up their long neck.

Hypertension affects the blood vessels as well as the heart. Chronically high blood pressure tends to induce microtears on the interior surface of the arteries. This damage initiates inflammatory responses that stiffen the arterial wall and leads to deposition of plaque that narrows the diameter of the artery. All arteries resist such microtears by slightly stretching with each pressure wave of blood. However, near the heart the giraffe's carotid arteries have a particularly high density of elastic fibers, giving them greater capacity to stretch rather than tear in the region of maximal blood pressure.[10] Farther up near the brain, the carotid arteries have fewer elastic fibers but more vascular muscle. This added muscle for dilating and constricting the arteries gives them added control for regulating the blood flow to the brain as the giraffe raises and lowers its head over 5 meters (16.5 feet) to drink water from the ground. Such movements can change blood pressure by nearly threefold. Giraffes clearly need some way to avoid the faintness humans feel when we rapidly change our head height by only about 1 meter (3 feet).

• • •

The high blood flow in the neck may well contribute to the human capacity for thought, but it also plays into our images of horror. For much of the twenty-first century, Americans have been infatuated with vampires, usually the young sexy types roaming their neighborhoods incognito, rather than the white-faced, high-collared denizens of Transylvania. Anne Rice's collection of novels *The Vampire Chronicles* (1976–2018) launched this contemporary wave of vampire obsession, selling more than a hundred million copies. The 1994 big-screen adaptation of *Interview with the Vampire* made over $225 million. The *Twilight* series of romance novels (2005–15) has sold over 160 million copies, and the corresponding films have grossed over $3 billion. Even the spoof-movie *Vampires Suck* (2010) opened number one and netted nearly $100 million at the box office.

Much has been written about what underlies this obsession. Vampires are alluring perhaps because they embody the tension between danger and erotica, passion and protection, exotica and intimacy, or mortality and eternal youth. Regardless, just beneath the surface of all vampire stories are the carotid arteries. Vampires would not live in our imagination if we were not deeply aware of the spouts they drink from. The carotids run up the neck unprotected and less than two centimeters (one inch) below the skin. In daily life they feed our most vital organ, the brain, and in our stories they feed the irrepressible appetite of vampires. While many people were adrift in the fiery passions of vampires, a small group of physics students at the University of Leicester wanted to get down to the cold, hard numbers of vampire behavior. Exactly how long would it take for a vampire to drink its meal? Publishing their work in the *Journal of Physics Special Topics*—it is indeed a special topic—the students presented their model incorporating a complex set of variables of carotid blood flow: blood pressure, blood density, vascular bifurcations, the pore size of the fangs, and so forth.[11] Here is their precise conclusion: 6.4 minutes to get a wine bottle's worth (750 milliliters) of warm blood. For comparison, when we donate our blood at a blood drive, our offering is half the volume and takes ten times longer.

While blood flow through the carotids elicits our darkest irrational fears, blood flow through other smaller, more superficial neck arteries broadcasts some of our most private emotions. When embarrassed or ashamed, many people involuntarily blush at the neck and face. Some people might even rather be attacked by a vampire than endure an unwanted episode of blushing. Blushing can be so cruel because it violates our privacy by announcing our shame or embarrassment. These emotions gleam from the most visible parts of the body: the face and the neck. Moreover, blushing amplifies itself. Embarrassment causes blushing, which causes more embarrassment. An awkward social exchange, an unwitting breach of etiquette, or even an unprovoked realization of guilt might excite involuntary fight-or-flight responses in our autonomic nervous system, which can include dilating arteries to the skin of the face and neck. This skin is particularly well vascularized and thin, so the increased blood flow shines forth. Often to our dismay, by diverting blood to the exposed skin of our neck, we expose our emotional vulnerabilities.[12]

Blood vessels are our vital connections between the heart and the mind, and the massive flow of blood through these tubes supports our human intelligence. But these vessels lie near the surface, exposing us to threats, both real and imagined, and eliciting our primitive emotions. Blood of life, but also blood of horror and embarrassment.

FOOD AND THE ESOPHAGUS

Unlike the flow of blood up the neck, the transit of food down the neck is intermittent rather than continuous. Swallowing is a brief, all-or-nothing event of ingestion before the long gradual process of digestion. With every bite and sip, the body must determine when and whether it will make the (almost) irreversible decision to finally gulp. Normally all goes well, and the body is given its needed nutrition. But occasionally the gulp goes the wrong direction, and we face the terror of choking. From our fish ancestors we inherited an awkward splicing of tubes in the neck that makes us choose between swallowing and breathing. A correct swallow is a matter of life and death.

The decision to swallow involves a series of sensors in the mouth, beginning with chemical sensors (taste buds) located mostly on the front of the tongue. If the food passes the taste test, it is shifted to the back of the mouth, where it is chewed and lubricated with saliva. On the back of the tongue, incredibly sensitive mechanical sensors detect the particle size and wetness of the masticated food and assess whether it is sufficiently soft and fluid to take the big plunge. If everything still passes the tests, we begin to swallow through a precisely timed and rapid series of contractions in more than a dozen neck and facial muscles. After this point it's all out of our voluntary control and reflexes take over. To successfully swallow, muscles in the mouth and throat must guide the food in the right direction. The tongue presses against the teeth and roof of the mouth to prevent food from coming forward, and the back of the throat elevates to block the passageway to the nose. The back of the tongue then pushes the food backward while the muscles of the upper throat squeeze the food downward.

Here at the top of the throat, the food is at a precarious junction. It could potentially veer into the wrong passage and wind up in the lungs rather than the stomach. To prevent such aspiration of the food into the trachea, we first close the vocal folds of the larynx with every swallow. Then comes the visible part—gulping. Everyone can now see our final decision to ingest because the larynx and the hyoid (the visible lump on the throat) elevate. The rising larynx causes a cartilaginous flap (the epiglottis) to fold down to occlude the trachea. After the epiglottis opens (in less than one tenth of a second), food enters the esophagus and moves toward the stomach. Almost always, we do this entire multistep, highly coordinated complex operation perfectly well without thinking. Often, we even combine swallowing with complicated actions such walking, driving, reading, or most any other activity—except breathing and talking. The neck cannot perform all its functions simultaneously. The confluence of the food and air tubes in the neck requires that swallowing precludes, if only momentarily, the universal act of breathing and the peculiarly human act of speech. It's one of evolution's trade-offs.

Because food and air have a common passage through the throat, every gulp brings with it a life-threatening possibility. If the epiglottis (the valve between the esophagus and trachea) does not work just right, a wrong swallow could occlude the trachea and steal our last breath. Choking kills about five thousand Americans annually and is the fourth leading cause of accidental death.[13] To Aristotle, the epiglottis allows us to negotiate an inherent geometric problem in our anatomy: "It is a contrivance of nature to remedy the vicious position of the windpipe in front of the esophagus, [and] that position is the result of necessity."[14] To convince yourself of the "viciousness" and "necessity" of the crossing tubes in the neck, try this exercise. First, press on the front of your throat to convince yourself that the hard trachea is at the front. Now,

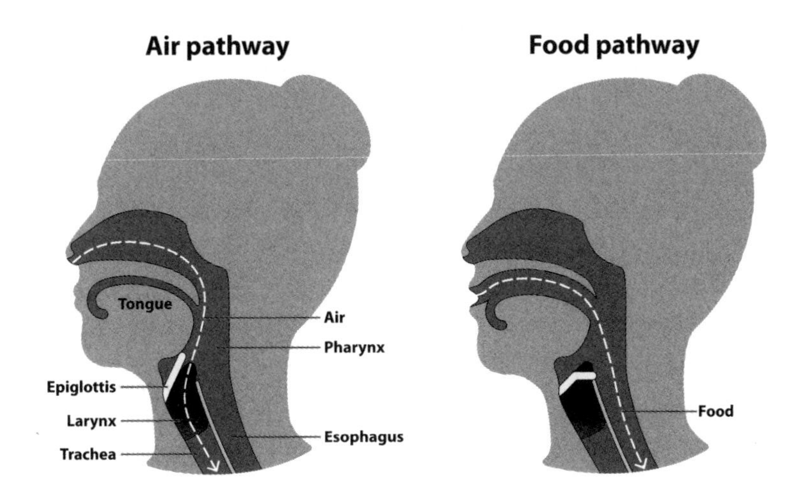

Figure 8. Pathways of air and food through the neck, showing the epiglottis as a valve that guides air into the trachea and food into the esophagus. Illustration by Netta Kasher, 2024.

place your index finger at your nose, the entry point of air, and your middle finger at your mouth, the entry point of food. Draw your hand downward and you will notice that your index finger now lies behind your middle finger rather than in the front, where the air needs to pass down the trachea. The only way to get the air to the front is to cross the tubes and guide the air forward down the trachea and the food back down the esophagus.[15]

This crisscross is the geometric necessity that Aristotle spoke of. The viciousness is that if the air and food are not directed into the correct pathway by the epiglottis, we choke. If our throats were engineered products, surely there would be a massive recall to fix this design. But for biologist Louis Held, the epiglottis is just one of the many "silly, stupid and dangerous" quirks in the human body. It illustrates "the sad fact . . . that evolution is no engineer. It's just a tinkerer. Worse than that, [it is] a *myopic* tinkerer."[16] Evolution by natural selection solves short-term problems from the local options available and resists going back to fix suboptimal designs. While Aristotle's argument stems from the geometry of the adult, Held offers an explanation based on the embryonic formation of the crossing tubes and the nearsighted action of evolution. The first tube formed in all vertebrate embryos is the gut tube. The upper end of this tube forms the esophagus, spanning from the mouth to the stomach. In the embryos of air-breathing vertebrates, a pouch buds off this tube toward the front to form the lungs. So the tube leading to the lungs (the trachea) branches forward and passes in front of the esophagus. That is, the forward position of the trachea arises from embryonic construction.

Logical, forward-thinking engineers would probably design a very different configuration, like a separate opening below the mouth to transmit air directly from the outside into the lungs, or perhaps somehow move the lungs behind the esophagus. But evolution settled for an adequate short-term solution, taking the mouth and esophagus already used for feeding, joining them to the novel trachea used for breathing, and separating one from the other by means of the epiglottis. At the time, when our first fish ancestors began gulping air at the water surface, this permitted them to draw in air through the mouth without inventing a whole new connection to their rudimentary breathing apparatus. But once this design was established, terrestrial vertebrates were stuck with it, and thereafter we all face the vulnerability of suffocating while eating.

All air-breathing vertebrates have some sort of intersection between the respiratory and digestive system. However, keeping flow through these tubes separate is especially challenging in adult humans. In other mammals and in infant humans the larynx and epiglottis are relatively high in the throat. The passage from the nose to the trachea is short and relatively direct. Food entering the throat largely takes a path around this air passage. In essence, they form a partial "tube within a tube" that functionally separates the airway and the foodway.[17] This minimizes the crosstalk between the pathways and enables infant humans to suckle at the breast and breath simultaneously.[18] However, as humans grow in the first months of life, our throat changes considerably. The larynx descends, and this downward migration has important advantages for expanding our vocal repertoire (as discussed in chapter 6), but it complicates our swallowing. The descending larynx drags the epiglottis and root of the tongue down with it and largely obliterates the tube-within-a tube configuration. With the larynx lower down, there is simply a larger shared space for food and air to mix. At only three months of age, we face a cruel compromise: with the descending larynx we gain our remarkable human capacity for vocalizing at the cost of a lifetime peril of choking. If humans had been designed by true innovators, we might have acquired separate tubes for eating and speaking. Alas, evolution is a renovator. It started with the original simple feeding tube and spliced in breathing and vocalizing. It seemed like a good idea—until we try to eat while we breathe or talk.

• • •

Because of the crossing paths of food and air in the throat, most air-breathing vertebrates must choose to close *either* the trachea *or* the esophagus. However, the largest animals to ever inhabit the planet must sometimes tightly close off *both* tubes at the same time. Baleen whales, such as blue whales and fin whales, acquire their huge amounts of food through a behavior called lunge feeding. When a whale encounters a swarm of krill (shrimplike crustaceans), it swings open its jaws at a right angle and engulfs an enormous quantity of water that balloons out its mouth and throat. This huge expansion is possible because the throat is built from a pleated, elastic layer of blubber and muscle. During engulfment, the throat expands by stretching like an accordion. Then, as the whale closes its jaws, it shuts *both* the esophagus and trachea, lifts the floor of the mouth and throat, and expels the water through fibrous plates (baleens) attached near the upper "lip" of the mouth. The krill is sieved out and stays behind in the whale's mouth. Finally, with the water gone, the whale opens the esophagus and gulps down the krill.

The masses and forces in such a feeding episode are staggering. Each lunge collects about 70,000 liters (roughly 18,000 gallons) of water weighing about 70,000 kilograms (150,000 pounds), approximately 50 percent more than the whale's body mass and about twice the mass of a loaded semitruck. Opening its mouth so wide creates a drag that almost halts its enormous body moving at about 3 meters per second (roughly 7 miles per hour).[19] Given these huge forces, how does the whale avoid choking on the water? Researchers discovered only recently that baleen whales have a large fleshy blob in their throats—an oral plug—that enables them to prevent water from flowing into either the esophagus or the respiratory tract during lunge feeding.[20] Led by Kelsey Gil, a research team recovered tissues from several fin whales and dissected the throat region. This small portion of the whale alone weighed several hundred kilograms and required a forklift to move into the dissection lab. They found a mass of fatty and muscular tissue at the back of the mouth that can completely occlude the pharynx, the shared space for the passage of both food and air.

Based on the orientation of the muscle fibers, the researchers concluded that the oral plug can be pulled upward to block the nasal passages while modifications to the larynx can be pressed against the opening of the trachea to block the passage to the lungs. None of these anatomical specializations has been found in any other animal, so it is likely they are part of the evolutionary remodeling that allows baleen whales to consume such copious prey and achieve such gargantuan sizes. Perhaps it is not coincidental that we call whales gargantuans. This word originated from Gargantua, a giant voracious character in the novels of François Rabelais. The name was allegedly derived from the Spanish/Portuguese word *garganta*, meaning "throat."

• • •

For most vertebrates the esophagus is a tube that simply transmits nutrition from the mouth to the stomach, where it begins the process of digestion and distribution to the rest of the body. But for some birds the esophagus has evolved to serve a broader array of functions related to nutrition. These evolutionary specializations occur in tandem with an important evolutionary loss: ancestral birds lost teeth. For nonavian vertebrates teeth play a crucial role in obtaining nutrition, piercing and gripping animal prey, or clipping and ripping bites of plants. In mammals the molars then grind it all up to aid in digestion. But birds have no teeth. Indeed, the avian lifestyle is apparently so incompatible with teeth that they lost them five separate times during their evolutionary history.[21] Moreover, birds have nothing very massive at all in their heads—no teeth, no heavy jaws, no powerful jaw muscles—which has major implications for their necks, as discussed previously.

The evolutionary loss of heavy chewing anatomy is usually interpreted as an adaptation for flight. The disadvantage of a heavy head extending forward from the center of mass is clear.[22] The lack of a chewing apparatus in birds means that all their food processing must occur after they swallow. Much of what occurs in our mouths occurs in their throats or lower down. In many bird species, after food is swallowed, it passes into a bulbous enlargement of the esophagus (the crop) that accomplishes the initial stages of digestion. The crop secretes acid to begin the chemical breakdown of food and mucus to lubricate the passage of food through the gut. In some species the food then travels to a muscular gizzard in their abdomen that contains pebbles to grind up food before it is fully digested chemically in the stomach. With their loss of teeth, birds made diverse evolutionary modifications of their esophagus and upper digestive tract that expanded their range of behaviors and lifestyles.

In some birds the crop functions as much for holding food as it does for passing it on down. In seed-eating birds such as chickens, the crop is especially well developed and serves as a temporary storage area. Such granivorous birds often forage in open spaces but retreat to protected spots to digest in private. Their crop allows them to swallow a lot of hard food in the field, store it away in their crop, run back to hiding, and then slowly pass their food into their stomach for digestion. Similarly, vultures and other scavengers gorge themselves at carcasses, where heavy competition with other scavengers favors a fast eater, then they fly away with a crop packed with carrion to digest. Competition, or rather outright thievery, also figures into the way many fish-eating birds use their crops. When seabirds (such as pelicans, cormorants, and boobies) are returning from a successful foraging trip at sea, they face the threat that less successful birds will snatch their catch from their beaks midair. By swallowing fish into the crop, they can protect their meal until they digest it on land or regurgitate it for their chicks.

Closer to home, another bird uses its crop to manufacture food, not just store it. Pigeons abound in urban settings in part because of their peculiar manner of feeding their chicks. As the ultimate opportunists, they scavenge whatever food is available in the urban landscape—berries or grass in the park, stale bread from the dumpster, or insects crawling on food wrappers. Pigeons breed whenever they get enough calories to make eggs. After the chicks hatch, it could be tricky to feed them because their improvised food sources are so diverse and unpredictable. So pigeons do what mammals do: they make their own food for their young. In a remarkable convergence between pigeons and mammals, their crop is analogous to the breast. It makes a whitish, high-fat, high-protein substance, loaded with antibodies. Although this substance is commonly called "pigeon milk," it is more the consistency of cottage cheese than milk. Like mammalian lactation, pigeon milk production activates just before it is needed by the young—two days before the eggs hatch—and shuts down after the chicks fledge.

Despite their similarities in function, the pigeon crop and the mammalian mammary gland are anatomically very different structures. Mammary glands are modified sweat glands that make milk, but birds do not have sweat glands at all. Instead, pigeons' "milk" is produced by modified skin cells (keratinocytes) in the crop. The entire fatty keratinocyte layer sloughs off to form the "milk," infusing it with all the nutritious molecules manufactured inside the cells. Curiously, this entirely distinct form of milk production is controlled by the same hormone (prolactin) that controls milk production by the mammary gland.

Moreover, male pigeons as well as females produce milk, and this "male lactation" is also controlled by prolactin.

Across vertebrates, diet influences the mode of locomotion. Ingestion of large quantities of calorie-poor foliage, such as leaves and grasses, is apparently incompatible with flight. Although many birds eat mostly calorie-rich plant products—for example, seeds and nectar—very few birds (only 3 percent of all avian species) include leaves as a substantial component of their forage. This diet would weigh them down. Leaves are composed primarily of cellulose, which cannot be broken down by vertebrate digestive enzymes. So most leaf-eating vertebrates (like deer, rabbits, and iguanas) have in their guts a fermentation chamber harboring cellulose-digesting bacteria. Even with the aid of these bacteria, digesting leaves requires a lot of time and water. Consequently, these fermentation chambers hold a large mass of partially digested slurry, and thus they are usually incompatible with flight. However, one bird, the (almost) flightless hoatzin from tropical South America, has evolved a large fermentation vat in their lower throat, enabling them to derive enough calories from a diet of leaves. This gives them a bad name—"stinking pheasant" in the local language. It's not a coincidence that hoatzins smell like manure, since they engage in the same basic process of cellulose digestion as cows. While cows use their rumen (a modification of their stomach), hoatzins use their crop (a modification of their esophagus).

The hoatzin's fermenting crop was a great surprise when it was first reported in 1989.[23] Virtually every other animal that uses its foregut for fermentation is a large and/or slow-moving mammal (ruminants, such as cows and giraffes; and kangaroos or sloths) that does not have great energy requirements. It seemed unlikely that leaf fermentation, a relatively inefficient digestive process, could ever provide enough energy for a small (less than 1 kilogram [2 pounds]) bird, since on a per gram basis small animals have higher caloric needs. Several morphological and behavioral adaptations allow hoatzins to beat the prediction. First, their crop has hard, interior ridges that grind up the ingested leaves, improving the digestive efficiency of fermentation. Second, hoatzins selectively forage on young leaves, which are more nutritious and contain a greater fraction of a more digestible form of cellulose, hemicellulose. Finally, they simply do not fly very much, so their energy demands are lower than predicted for birds of their size. This reduction in flight is not just an energy-saving behavior; it is also likely a direct consequence of the expanded crop. In hoatzins the crop comprises about 70 percent of the entire digestive tract and weighs about 17 percent of the entire weight of the bird. This huge expansion of the crop has led to a reduction in the adjacent breastbone, where the flight muscles attach. Consequently, even if they had calories to spare, they could not fly very well. Their throats get in the way.

From small birds to gargantuan whales, the esophagus has diverse shapes and functions. But fundamentally it is a tube serving the basic need to move food internally. Both evolutionarily and embryologically, the esophagus is part of our oldest interior space. In our earliest vertebrate ancestors and at our earliest developmental stages, the gut tube forms first. At one end is the apparatus for apprehending and macerating food; at the other end is the vat that chemically breaks it down and absorbs it. The connector in between—the throat and esophagus—acts both as a crucial gate and as a transporter of nutrition. But later in evolution and embryology, the gut tube had to accommodate a shared space with the air tube. This new addition opened up the wide world of terrestrial existence, but splicing it into the more ancient tube with near-sighted tinkering left us with a lingering flaw: a misdirected morsel can be deadly. We have to get our swallow right every time.

AIR AND THE TRACHEA

The food we eat is the fuel for our survival, but, as we know from campfires, fuel without air barely burns. Food molecules by themselves have relatively little usable energy unless they are combined with oxygen in cellular metabolism. To ensure oxygen is distributed to all our cells, we transport air from the nose and mouth via the trachea to the lungs, where oxygen diffuses into the blood and circulates around the body. Our breath allows us to burn brighter.

Air flow in the trachea oscillates: inhalation and exhalation, back and forth twelve to eighteen times per minute, about twenty thousand times a day, with air flowing at a calm rate of 8 kilometers (about 5 miles) per hour. It is one of the most predictable and regular rhythms of life. But occasionally, all hell breaks loose. With a tickle or irritation in the throat, you reflexively shut the epiglottis, narrow the tracheal diameter, vigorously contract your diaphragm and rib muscles to build up pressure in the lower airway, and then open the epiglottis to release the pressure in a violent exhalation of air moving up to 1,000 kilometers (625 miles) per hour.[24] In other words, you've just coughed. If the coughing continues day after day, it may send you to the doctor. Indeed, of all our ailments, a persistent cough is the symptom that most commonly prompts people to visit their primary care physician. Every life-giving breath brings with it unwanted particles—dust and pathogens—that could easily damage the lungs. Coughing gives us a way to eliminate these inevitable contaminants.

The trachea and bronchi of the lungs secrete a sticky mucus that traps the particles and is then swept upward by microscopic cilia that line the airway. This mucus is our guardian goo. But from the standpoint of elimination, the stickiness is a problem. Mucus adheres tightly to the walls of our airway, and if we want to do a mass purge through a cough, it requires a considerable shearing force of air moving upward to cause the mucus to break away from the surface. In a healthy person with a normal mucus layer, a cough is sufficient to dislodge the mucus. But in certain disease states with highly concentrated mucus, such as cystic fibrosis, not even a 600-miles-per-hour whirlwind is strong enough to overcome the tighter adhesion of the mucus to the lining of the air passages. The mucus dangerously accumulates in the respiratory tract despite heaving coughs.

During the COVID-19 pandemic there was nothing quite so alarming as a stranger coughing in a crowded public space. Everyone knew that COVID strikes first in the throat and that airborne particles cast through the throat were the primary means of transmission. The ability of the virus to make its victims cough was no doubt key to its success as a pathogen. Yet everyone was also aware that people cough for other reasons—for example, allergies, excess mucus production, and non-COVID pathogens. From the time of highest COVID contagion, I recall two sorts of facial expressions when someone coughed in public. Most people gave a restrained grimace, a combination of fear and disgust. But the cougher often had an ashamed look: "Honest, I've been super-careful, and I just tested negative yesterday."

Much of this fear and shame could have been avoided if we had some way to distinguish a contagious versus harmless cough by its sound. Some people claim they can: "That person sounds really sick." Such ability to discriminate between a dangerous and harmless cough would be highly adaptive, not just during a pandemic, but throughout evolutionary history. But research published in the first few weeks of the pandemic (April 2020) showed we cannot assess the danger of a cough by simply listening. Researchers at the University of Michigan and the University of California, Irvine, presented subjects with forty coughing sounds, half of which were recorded from people known to have an infectious disease, and the other half from people known to be non-infectious. Contrary to folkloric claims, the researchers found no evidence that people could correctly identify the infectious status of the cougher based on sound. So, although it makes perfect evolutionary sense for us to make such accurate discriminations, it is not something we can do, at least not innately.

However, perhaps we could learn to make these distinctions if we had enough data. In the early months of the pandemic, Brian Subirana and colleagues at the Massachusetts Institute of Technology launched a website where people from all over the world could submit recordings of their cough made on their cell phones and answer questions about their COVID status.[25] All told, about thirty thousand people uploaded their coughs, including roughly twenty-six hundred who were COVID-positive. With this huge dataset, Subirana's group trained computers to discriminate between the coughs of the COVID-positive patients and an equal number of matched COVID-negative controls. This machine-learning approach generated an algorithm that was subsequently tested for its ability to identify infection status based only on the sound of coughs. Even though humans could not discriminate between COVID-positive and -negative coughs by ear, the algorithm was almost perfect. It accurately identified 98.5 percent of the sample that independently tested COVID-positive through nasal swabs.

Based in part on this research, the Australian company ResApp Health developed a cell phone app that could diagnose infectious status remotely and quickly based on cough sounds, without swab-based tests that required more time and expense. Early clinical trials of the app showed a similarly high (92 percent) detection rate in real-world settings.[26] The hope is that related technology will be able to detect all sorts of respiratory diseases (asthma, pneumonia, chronic obstructive pulmonary disease) by sound.[27] If technology continues in this direction, it seems possible we might all be coughing into our cell phones as often as we visit the doctor, take a swab test, or even enter a crowded public space.

• • •

Coughing, with our breath moving at 600 miles per hour, can feel disturbing if not violent. But perhaps even more terrifying is the opposite—the absence of breath. Yet a remarkably high number of people find themselves dangerously out of breath every night. As we move through the world during the day, upright and vocal, the design of our head and neck seems like a good idea. We have a broad panorama and an agile voice for communicating. But when we get horizontal to sleep, the design of our neck and vocal apparatus becomes questionable. Just after drifting into slumber, some people enter a foreboding cycle of breath and breathlessness: sleep apnea. During sleep the regular periodic breathing is interrupted when the tissues near the voice box collapse down on the flow of air, and, for a period of up to a minute, the sleeper verges on suffocation. Blood oxygen content starts to fall, and blood pressure starts to rise. The heartbeat may even stop for up to ten seconds. Soon the brain detects the urgency of the situation and rouses the body enough to take a big corrective gasp. Then the cycle starts over, repeating as many as thirty times an hour.

Usually the breathless episodes are not consciously registered by the sleeper, but over time apnea can have serious health impacts throughout the body. Below the neck, untreated apnea can lead to chronic hypertension and kidney malfunction. Above the neck nightly disruption of oxygen delivery to the brain increases the risk of mood disorders and even dementia. Sleep apnea is remarkably prevalent. By one recent estimate, about one billion people worldwide—about one in seven—suffer from apnea.[28] It is particularly common in older people, whose throat tissues tend to lose muscle tone and elasticity, and in obese people, whose throats are more collapsible because they must support more tissue mass. Apnea is a nearly unique human phenomenon. Flat-faced dogs, such as English bulldogs and pugs, are the only other animals known to display apnea.

Why are humans so uniquely prone to this nightly brush with suffocation? Put simply, the same bipedal posture and vocal anatomy that give humans so many advantages also impose space limitations in the throat that make it more vulnerable to collapse when we sleep.[29] In our upright stance, the head sits directly atop the spine and, along with our flat faces and short necks, this leaves relatively little space between the chin and the thorax. Moreover, the passage for air from the horizontal nose and mouth cavities to the vertical trachea takes a right-angle turn in the throat. In humans the epiglottis is positioned below this bend, and compared to other mammals, it is located farther down this passageway. This lower position leaves the upper throat relatively unsupported, distensible, and apt to collapse. In addition, over human evolution, the larynx (voice box) also descended down the throat and the airway narrowed, which enhanced our vocal prowess (see chapter 6). The tongue, which in other animals is flat and confined to the mouth, bends and extends down the throat in humans where it can subtly modulate the sounds generated by the larynx for speech. However, in this position it can also press on the deeper airway of the throat. In short, the throat became both more crowded and floppier during evolution. When we lie down and drift into sleep, the same features that grant us our upright and verbal life tend to threaten the breath of life. It is yet another of evolution's harsh trade-offs.

• • •

In 2020, Paolo Macchiarini was indicted by Swedish courts on three acts of aggravated assault. Three people died after Macchiarini took his knife to their throats, the indictment claimed. The prosecutor wrote, "The acts are to be judged as grave because Paolo Macchiarini has inflicted severe bodily injuries and very severe suffering.... In addition, [he] has shown particular ruthlessness and callousness."[30] Terrifying, yes, but probably not what you are thinking. Macchiarini cut the throat of these three people in the name of medicine. Working at Sweden's famed Karolinska Institute, he was attempting to surgically treat patients with malfunctioning tracheas through a novel and potentially groundbreaking transplantation procedure. Tracheal transplantation had long been a "holy grail" of reconstructive surgery. In 2008, Macchiarini began implementing a technique in which the trachea from a donor was "decellularized" into a tracheal scaffolding and reseeded with the stem cells of the patient.[31] Later, he attempted a similar procedure but instead started with a synthetic trachea.[32]

In principle, these techniques would solve the perennial problem of immune rejection by the patient because the only living tissues implanted were the patient's own. At the time, the procedure was so promising that the Karolinska Institute recruited Macchiarini in the hope that he might even "bring the Nobel back" to Sweden.

Macchiarini ascended to near-celebrity status. His first major transplantation operation at the Karolinska Institute was covered on the front page of the *New York Times*.[33] But by 2016 it all started to unravel into a scandal of fraud and deception. Of the nine patients Macchiarini operated on from 2011 through 2014, eight died. The reconstructed tracheas did not integrate well with other tissues and did not effectively clear mucus from the throat. It came to light that the procedure had not gone through routine testing (such as prior trials in animals) or received approval from ethical review boards. Some of the surgeries were conducted without prior patient consent. To sustain support for his work, Macchiarini greatly overstated the success of the procedures in interviews and published research articles.[34] In 2022 a Swedish jury found Macchiarini criminally liable for his tracheal procedure on a Turkish woman that led to her three-year stay in intensive care at a Swedish hospital and her eventual death.

While Macchiarini's epic story is full of hubris and tragedy, the trachea itself is not terribly dramatic. It is mostly just a pipe that, compared to its neighbors (the larynx above, the esophagus behind, or the neck muscles to the side), does not do a lot. It transmits air to the lungs, secretes and sweeps out particle-trapping mucus, and not much else. Given this relatively simple function, it is surprising that the development of tracheal transplantation was delayed so long compared to transplantation of organs that serve much more complex functions. Kidneys were first transplanted in 1954; liver, heart, and pancreas in the late 1960s; and lungs and intestines in the 1980s. Plumbers have been splicing in pipes for millennia, but doctors have only very recently developed the ability to splice in a replacement windpipe. The difficulty in tracheal transplantation lies not so much in the pipe itself but in the vasculature that feeds the pipe. Unlike other organs that receive blood through only a few large arteries, the trachea receives its blood from a complex network of vessels. As the two main feeder arteries approach the trachea, they branch into minute arteries that enter the side between each of the eighteen cartilaginous rings and travel around the circumference of the trachea. For decades it was believed that each of these small arteries must be preserved for the cells in each tracheal segment to survive. Based on this ornate vascular pattern, hundreds of clinical articles dismissed the possibility of a full-scale transplant.

After the Macchiarini debacle, a team of doctors and researchers at Mount Sinai Hospital in New York led by Eric Genden initiated a new approach for preserving the blood flow in the tracheal transplant. In 2021 they attempted a procedure based in part on modern surgical technologies and in part on careful rereading of old anatomical descriptions of tracheal vasculature, some dating back to the 1800s. Certain often overlooked vessels feeding the trachea first pass through other nearby organs, the thyroid gland and esophagus, before they branch into the trachea. With this rediscovered vascular anatomy in mind, Genden and his team transplanted large portions of the donor's thyroid and esophagus along with the trachea, enabling them to retain some of the intact downstream vasculature to the trachea. The surgery involved more than fifty professionals and took eighteen hours with a lot of very fine microvascular suturing. With the help of nineteenth-century anatomists and honest surgeons, it appears that the patient is thriving.[35]

• • •

Animals never face the challenges of tracheal transplant surgery, but some animals, particularly those with long necks like giraffes, face another tracheal problem, a "dead air" problem. The difficulty of transporting air through a long trachea becomes clear when you consider an artificial tube we humans sometimes use for breathing: a snorkel. First, notice that all snorkels are quite short, less than 0.3 meter (1 foot). Then, imagine trying to breathe through a very long, say, 2-meter (6-foot) snorkel. It's impossible. The reason is that the space inside a long snorkel holds a lot of "dead air"—air that moves back and forth inside the snorkel but never exits at either end. To get fresh, oxygen-rich air into your lungs, you also have to suck all the stagnant air that occupies the tube. You could conceivably reduce this problem by making the snorkel narrow in diameter, but this would cause a lot of resistance from air friction along the walls of the tube. That is, if you made the snorkel the diameter of a drinking straw, you would draw in less stagnant air, but it would take a huge sucking power to get the fresh air to your mouth. A long-necked animal like a giraffe faces the same problem with its trachea.

The problem is even greater in a creature whose neck is many times longer than a giraffe neck. Certain sauropod dinosaurs—the largest terrestrial animals to ever have lived—had necks that were up to 15 meters (50 feet) in length, six times longer than the world-record giraffe neck.[36] The closest living relatives of sauropods (birds) also tend to have long necks. Birds solve the problem created by their necks with a curious respiratory anatomy that is very different from ours. They breathe in a flow-through manner, rather than the back-and-forth pattern of most terrestrial vertebrates. Alongside their lungs birds have two sets of air sacs that act as bellows and air-holding reservoirs. The path of the breath through the respiratory system is complicated but, in essence, it is a loop.[37] This unidirectional flow of air across the lungs is more efficient than the back-and-forth pump in mammals, and it minimizes the dead air problem. Fossil evidence indicates that this avian flow-through respiratory system was present in sauropods, and it likely contributed to the sauropod's ability to breathe through their very long tracheas.

Questions of respiration also have bearing on whether sauropods dwelled on land or in water. Nineteenth- and early twentieth-century biologists thought that an animal this size could never have supported itself on land. Moreover, their large size raised thermoregulatory issues. Very large animals, at least those living in warm climates, face the risk of accumulating heat. They have relatively low surface area for dissipating heat in relation to the mass of their tissues, which generate heat. This thermoregulatory issue led some paleontologists to postulate that sauropods lived in water to keep cool, using their long neck to keep their head up to see, eat, and breathe in air while their bodies were deeply submerged in cool water. Their aquatic surroundings would also help them support their very long heavy neck. By the 1970s this interpretation of their lifestyle fell out of favor, in part due to yet another consideration of the respiratory system. Placing the large body underwater would have caused such high pressure on the thorax that it would have been difficult for them to expand their lungs for breathing. Since then, paleontologists have reassessed the structure of sauropod bones and the structural requirements for supporting their body mass, and presently most believe that sauropods were largely, if not exclusively, terrestrial.

These three questions—where sauropods lived, how they breathed, and how they kept cool—are addressed in part by the specializations of the neck. Like modern birds and dinosaurs in other lineages, sauropods had an elaborate system of air-filled spaces within the cervical vertebrae. These sinus spaces were extensions from the air sac system and likely further enhanced the efficiency of their flow-through respiratory system. These air-filled cavities also reduced the weight of the neck and, along with sauropods' relatively small head, made it possible to support their head cantilevered at the end of a very long neck while roaming around on land. Finally, the necks of sauropods likely acted as radiators to dissipate heat. When they exhaled, hot air from the lungs passed through a long, moist trachea, releasing excess heat through the breath as well as the surface of the neck. These adaptations in the sauropod respiratory system and cervical vertebrae that supported their gigantic size are also found in modern birds and contribute to the light weight and high oxygen intake required for flight. In addition, they were present in the dinosaurian ancestors of birds even before birds took to flight. Thus it appears that anatomical features of the common ancestor to sauropods and birds were pre-adaptations that permitted two highly divergent paths: one toward the lumbering, terrestrial giants and the other to the small, aerial flyers.

• • •

Feel the pulse, breathe deeply, and take a sip. This—and not much else—happened on the day you were born. And just this was enough to convince the doctor and your parents that you were full of life. Carotid arteries are literally the lifelines from the heart to the mind. The trachea conveys the breath of life, both inspiration and exhaustion. The esophagus is the passage for sustenance. The neck serves heroically, often unrecognized, as a major conduit of life, a bridge daily transmitting vital fluids—literally a ton of them—between the head and the torso. Evolutionary modifications in these tubes have contributed to our uniquely human intelligence and vocal abilities. Ingenious adaptations in these tubes in our vertebrate cousins have enabled them to grow to gargantuan size, reach towering heights, and soar through the air.

As stalwart and adaptable as these structures are, they are also fallible and plagued with trade-offs. They evolved through a history of short-sighted, creative improvisation that has given us great capacities but also collateral vulnerabilities. The same neck that is flexible enough to bend the head is thin enough to be easily punctured. The same neck that can swallow is the same one that can choke. The same neck that speaks so well is the same one subject to nightly breathlessness. These are the glories and flaws of our quirky construction.

CHAPTER FIVE

Pace & Scaffolding

Hormones of the Neck

When Russia invaded Ukraine in February 2022, the whole world went on alert. Ukrainians took to arms, governments worldwide condemned the invasion and imposed sanctions, and corporations and cultural organizations suspended activities with Russia. Within weeks the threat became global as Vladimir Putin rattled his nuclear saber and his armies assaulted the nuclear power plant in Zaporizhzhia. This was not the first time the region had been on high nuclear alert; the worst nuclear accident in history had occurred thirty-six years previously at Chernobyl in northern Ukraine.

With threats and memories of nuclear disaster in the air, many Europeans sought protection in pills rather than in bomb shelters.[1] Pharmacies were overrun with people hoping to obtain tablets of a very simple chemical: iodine. In Belgium alone, nearly thirty thousand people requested iodine from pharmacies following Putin's comments that he was putting his nuclear forces on high alert. The going price for iodine tablets more than doubled, and still many pharmacies completely ran out of stock.[2] Factories in Romania revved up their production to thirty million iodine tablets, and the government announced it would distribute them to all their citizens under forty years old.[3] The concern was for their throats—in particular, the vulnerability of their thyroid glands to cancer from radioactive emissions. Iodine tablets would be their shield.

Nuclear emergencies commonly release radioactive iodine into the air and water, which can be absorbed directly into the body or indirectly through the meat and milk of domestic animals. Although other radioactive molecules can also be taken into the body, iodine is unique in that it concentrates in one specific tissue, the thyroid gland, located midway down the front of the neck. Natural iodine in the diet is an essential component of the hormones made by the thyroid gland, and consequently it is absorbed ravenously from the blood into the gland. When people are exposed to radioactive iodine, it potently accumulates in the thyroid, where the radiation can eventually induce cancerous mutations. In the decades following the 1986 Chernobyl disaster, an estimated four thousand people who were children and adolescents at the time of the accident have since developed thyroid cancer, directly attributable to the release of radioactive iodine.[4]

When nuclear threats rise, health officials often supply iodine tablets to the public as a protection against radioactive iodine. A large intake of nonradioactive iodine sates the thyroid gland's "hunger" for iodine, thereby minimizing the amount of radioactive iodine it absorbs. Such prophylactic distribution of iodine rolls out with each wave of nuclear threats: at the height of the Cold War nuclear arms races (1980s), in weeks following 9/11 in the United States (2001), and in the aftermath of the nuclear power plant leaks at Chernobyl (1986) and Fukushima, Japan (2011). With nuclear technology we ultimately have the capacity to exterminate all human life, but the most likely threat is to our necks.

EVOLUTION AND FUNCTION OF HORMONES FROM THE NECK

Outside of news reports on nuclear threats, about the only place we encounter mention of iodine is on a box of table salt. Just under the umbrella-toting Morton Salt girl, it reads: "This salt provides iodide, a necessary nutrient." When we salt our food at practically every meal, we take in tiny amounts of this trace nutrient, ensuring that we can always make thyroid hormones. Iodine does not always occur naturally with salt, but because it is an essential element in thyroid hormones, it is added to table salt at the advice of public health officials to keep our thyroid glands healthy. Such widespread distribution of iodine through salt has been one of the most successful of all global public health campaigns, greatly reducing the incidence of thyroid disease throughout the world.[5] However, the quest for sufficient iodine began far before public health programs. It started from the very earliest days of life on earth.

The most ancestral lifeforms, bacteria and blue-green algae, absorb iodine from ocean water. In these unicellular organisms and continuing into multicellular life, iodine acted as a powerful antioxidant that was crucial as organisms began using oxygen to fuel their cellular machinery. In addition, iodine can catalyze many biochemical processes because it is an especially unstable and reactive element. Primitive multicellular animals absorbed iodine from their diets, so higher iodine levels signaled more food. Thus, from very early on, iodine was likely linked to the energetic state of the organism. For marine organisms, obtaining iodine is not a problem because iodine is plentiful in seawater. However, as organisms colonized freshwater and terrestrial habitats, where iodine is scarcer, they evolved organs that concentrated it from their food and stored it in their bodies. Thereafter, this dietary molecule became incorporated into the internal physiology of the animal as hormones (such as thyroxine and triiodothyronine) for regulating the production of cellular energy and for coordinating major developmental changes.[6]

Another element necessary for all life is calcium. In unicellular and soft-bodied organisms, calcium commonly activates many intracellular processes and regulates cellular excitability. But in hard-bodied organisms, including vertebrates, calcium has an additional essential role because it is a main component of the skeleton—for example, the shells of shellfish and the bones of vertebrates. Calcium, like iodine, is abundant in the marine environment, and all sea animals absorb or pump external calcium into their bodies. As vertebrates evolved onto land, they left behind this ambient source of calcium and instead became reliant on their internal stores of calcium in their bones. This internal shuttling of calcium between bones and the blood came under the regulation of two other hormones secreted from the neck: parathyroid hormone and calcitonin.

In terrestrial vertebrates these three hormones—thyroid hormone, parathyroid hormone, and calcitonin—are secreted in minute quantities from glands in the neck and affect virtually every cell in the body. They regulate physiological processes fundamental to the human condition as highly energetic, warm-blooded mammals with heavy skeletons that support our terrestrial existence. These hormones regulate the pace and the scaffolding of our lives.

Hormones travel all over the body through the bloodstream, so in principle the glands that produce these hormones could be located anywhere in the body. But it is probably not coincidental that they reside in the neck. In our early ancestors, tissues that wound up in our neck played a crucial role in importing iodine and calcium just inside the mouth. To understand why these hormones are produced in the neck, we have to consider the evolutionary origins of their glands.

As soft tissues, endocrine glands do not fossilize. However, the peculiar iodine-concentrating feature of the thyroid gland allows anatomists to trace its evolutionary origins to the most distant cousins in our phylum, Chordata.[7] One of the most primitive living chordates, the lancelet, has a structure that bulges down from the floor of its oral cavity and concentrates iodine. This structure, the endostyle, has a drastically different function in these small (about 2–8 centimeters, or 1–3 inches) filter-feeding animals; it manufactures mucus that traps minute food particles just inside the mouth as water flows over the gills. So, from the very beginning of our chordate lineage, the primordial thyroid was located in the throat. In a closer relative, the lamprey (a primitive eel-like vertebrate), the endostyle forms and functions in the larval stage in the same way as it does in primitive protochordates, like the lancelet. When the larva undergoes metamorphosis, however, the endostyle closes off from the oral cavity and starts making thyroid hormone that circulates throughout the body. So the changes within the lamprey's lifetime replay the deep evolutionary transition of the thyroid gland from a feeding function to an endocrine function.

Unlike the thyroid gland, the parathyroid gland does not concentrate its own signature molecule, so its evolutionary origin has been more difficult to trace. However, recent studies of parathyroid genes suggest that the function of protoparathyroid tissue in fish explains the placement of the true parathyroid in the neck of land-dwelling vertebrates. Fish do not have true parathyroid glands. They have little need to tap into their bone calcium because calcium is abundant in their marine environment and readily transported into the blood through the gills. Moreover, because fish are buoyant in water, they do not need a robust skeleton to support their body. With these considerations, biologists speculate that the parathyroid glands arose in tandem with the emergence of vertebrates onto land, where calcium is less available and skeletons must be stronger.

Although fish have no parathyroid gland, they curiously have tissue that expresses the same gene, *Gcm-2*, that directs the development of the parathyroid gland in terrestrial vertebrates. In mammals and birds, *Gcm-2* is expressed in the developing throat of the embryo and is required for the formation of the parathyroid gland. In fish, it is expressed in tissue of the developing gills and is required for gill formation. Fish gills express two additional genes, *PTH* and *CasR*, that in terrestrial vertebrates play a central role in the hormone-producing and calcium-sensing functions of the parathyroid.[8] Thus it appears that in our fish ancestors, parathyroid-like tissue situated at the gills locally regulated the influx of calcium from the water to the blood via the gills. When vertebrates evolved onto land, this tissue transformed into a true parathyroid gland that, by producing a hormone acting on bones, regulated the flux of calcium from the skeleton to the blood. There is no need for this gland to be in our necks; it's there because it was in an analogous position in the gills of our fish ancestors.

• • •

The endocrine tissue of the neck is located just under the skin on the surface of the larynx (voice box). In humans it's about the shape and size (5 centimeters, or 2 inches) of a butterfly and looks like one single mass. But actually it is an amalgam of three separate hormone-producing tissues fused together into a single blob. Most of the tissue—the thyroid gland proper—produces thyroid hormones. Parathyroid hormone is produced by its own distinct tissue, the parathyroid glands, that reside as four separate, lentil-sized lobes dotting each "butterfly wing" of the thyroid. Calcitonin is produced by cells that are completely integrated and dispersed among the cells of the thyroid gland.

In humans and other warm-blooded animals, thyroid hormones regulate the rate of cellular reactions across the body, the overall resting metabolic rate. By controlling our baseline energy expenditures, they also influence our body temperature, activity, and weight balance. With excessive thyroid hormone production (*hyper*thyroidism), our furnace runs too high. We might sweat a lot, become irritable and restless, have an increased heart rate, and, unless we eat more, we lose weight. With low thyroid hormone secretion (*hypo*thyroidism), our body's furnace runs too low. We become cold, lethargic, and, unless we eat less, we gain weight. Without enough iodine in our diet, we cannot make adequate quantities of thyroid hormone, so the body reacts by stimulating the growth of the thyroid gland, resulting in a large swelling on the throat known as goiter. A parallel function of thyroid hormone is its role in embryonic development. It acts to coordinate the formation of many tissues but particularly the brain. Without sufficient thyroid hormone during fetal life, a child may have stunted growth and impaired cognitive abilities, a condition termed congenital iodine deficiency.

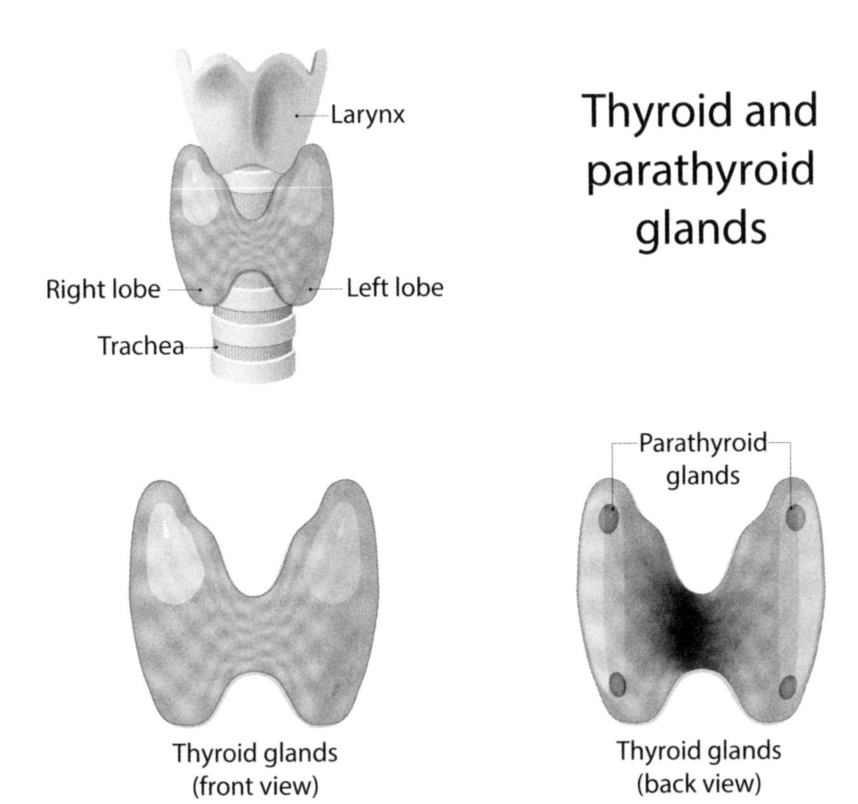

Figure 9. The human thyroid and parathyroid glands positioned on the front of the trachea. Illustration from iStock.

The other two hormones of the neck—parathyroid hormone and calcitonin—regulate the shuttling of calcium between the bones and blood. Since all the body's cells are highly sensitive to calcium, these hormones must tightly stabilize calcium levels in the blood. The bones have the largest mass of calcium in the body, so they are the primary internal source and repository of calcium. If blood calcium levels become too low, rising parathyroid hormone levels activate certain bone cells that break down bone, releasing calcium into the blood. In many vertebrates calcitonin serves to decrease blood calcium by inhibiting bone breakdown, although this effect is minor in humans.[9]

Disorders of these two calcium-regulating hormones are rare, but tumors on the parathyroid gland cause it to secrete excess parathyroid hormone, which leads to frail bones and overall weakness. In addition, because calcium is an important signaling molecule regulating electrical activity within cells, patients without enough parathyroid hormone become overexcited and may experience seizures, muscle spasms, and a racing heart. Parathyroid hormone fortifies our bones and moderates our excitability.

IODINE AND THYROID HORMONE IN HUMAN HISTORY

Over the past two hundred years, governments and other organizations have instituted salt iodination programs across the globe that have largely remedied regional patterns of iodine deficiency. But for most of human history this was not the case. People living in coastal areas, where iodine levels in the soil are high, have always consumed more than enough iodine from the plants and animals raised nearby. However, people in certain inland regions, where the soils are iodine poor, commonly experienced enlargement of the thyroid gland, goiter, and other thyroid disorders. "Goiter belts" ran through the Himalayas, Alps, and Andes but also in bands of lowland areas in interior Africa, North America, and South America. Historically, the prevalence of goiter and congenital iodine deficiency syndrome in these belts has been astonishingly high. An 1800 census ordered by Napoleon reported that four thousand inhabitants—almost 6 percent—of the canton Valais in the Alps displayed congenital iodine deficiency syndrome. In the US goiter belt during the early 1900s, 25–70 percent of children had clinical manifestations of goiter.[10]

Art and literature from these regions reveal that goiter was endemic for centuries and even millennia. A two-thousand-year-old tobacco pipe in the form of a human figure excavated in the Ohio Valley—right in the heart of the iodine-deficient zone of the United States—shows a

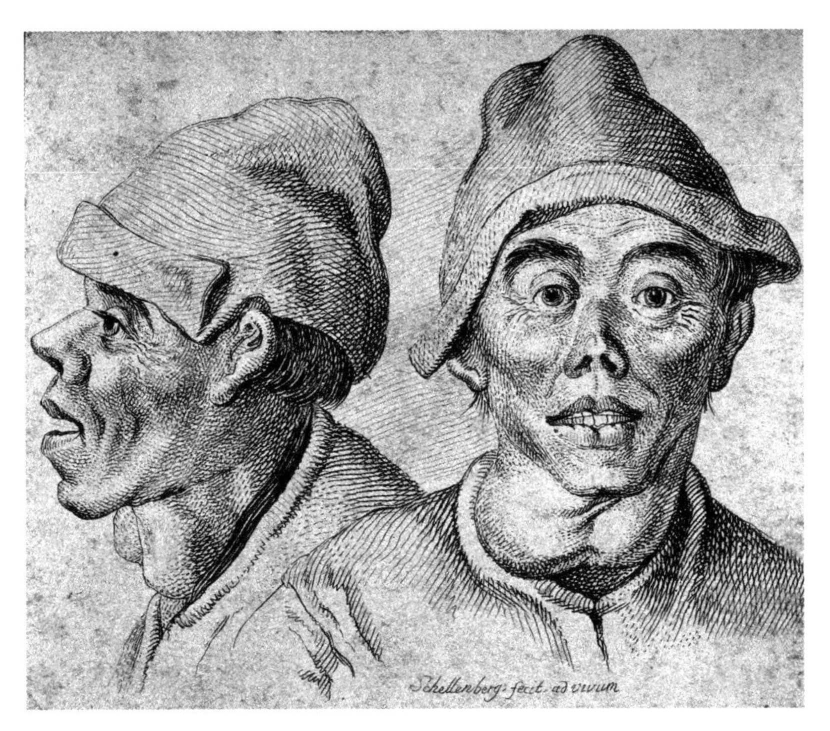

Figure 10. Etching of a man with goiter, by Daniel Friedrich Schellenberg, 1778, the Wellcome Collection (Wikimedia).

prominent bulge in the neck suggesting goiter. Renaissance-era paintings by Leonardo, Sanzio, and Caravaggio from Italy, near the goiter belt of the Alps, portray figures with noticeably protruding neck masses. In a survey of six hundred paintings and sculptures of the Italian Renaissance, over eighty works showed swellings at the throat.[11] In some cases these works seemed to emphasize the grotesque realism of the malformation, while in others it appears that a slightly swollen neck was a mark of feminine beauty.[12] Farther north, Shakespeare was evidently aware of the goiter endemic to the mountains of central Europe, even though he likely never visited there. In *The Tempest* he refers to the "mountaineers ... whose throats had hanging at 'em Wallets of flesh?" Centuries later, Mark Twain witnessed goiter firsthand through his travels to the Alps. In *A Tramp Abroad* (1880), he bids farewell to Europe with the words, "I am satisfied. I have seen the principal features of Swiss scenery—Mont Blanc and goiter—and now for home."[13] Because of the patchy distribution of iodine, certain peoples have contended with the scourge of endemic disease in their throat, and for ages this struggle has seeped into our cultural creations.

The notion that goiter had some sort of dietary cause dates back to ancient times. As early as 2700 BCE, Chinese medical practitioners treated goiter by feeding patients seaweed and "burnt sea sponge," both of which were shown in modern times to have especially high levels of iodine.[14] In the pre-Columbian Andes, highland communities with high incidence of goiter sought relief by trading their crops for seaweed and other seafood with coastal communities. These early trade routes between highland and coastal regions for distributing "thyroid medicine" established commercial connections that were formative in the integration of Incan civilization.

While the link between dietary seaweed and goiter had been known across the globe for thousands of years, the causal connection to iodine entered Western medicine in the early nineteenth century. The discovery came largely through the work of three French chemists.[15] In 1811 chemist B. Courtois observed a purple vapor rising from the ashes of seaweed treated with sulfuric acid, a process used in the making of gunpowder. He named the substance iodine, derived from the Greek word *ioeides* meaning "purple appearance." Another French chemist, J.B. Boussingault, working in Colombia, demonstrated that salt with naturally occurring iodine could protect whole populations against goiter. He measured iodine levels in the salt deposits around the Andes and found that communities with salt low in iodine had high incidence of goiter. However, when these communities imported salt from nearby goiter-free regions, their rates of goiter fell. These observations drove him and A. Chatin, a third French chemist working in goitrous regions of France, to advocate for the widespread iodination of salt to prevent goiter. By the early 1920s, government-sponsored iodine supplementation programs were established in the goiter belts of Switzerland and the Great Lakes region of the United States.

In the beginning, iodination programs were not always welcome. When iodized salt was introduced in Michigan in 1924, protests broke out for fear that iodine was a "poison" and might even *cause* hyperthyroidism. The first Bureau of Chemistry at the US Department of Agriculture initially attempted to mandate that packages of iodized salt be labeled with a skull and crossbones. To counter such fears, public health officials launched campaigns to convince skeptical populations.[16] South Carolina boldly proclaimed itself as "The Iodine State" and advertised the high iodine content of its agricultural products. Even moonshiners joined the campaign. Touting the natural abundance of iodine in their local concoction, the Hell Hole brand of "liquid corn" was sold proudly with the slogan "Not a Goiter in a Gallon."

Thanks to the systematic adoption of salt iodination in the twentieth century, endemic goiter and iodine deficiency syndrome have been practically eliminated in many previously plagued regions of the world. Mandatory iodination programs have been legislated in 123 countries, and worldwide almost 90 percent of households now consume salt with iodine.[17] Still, iodine deficiency remains a major public health issue, particularly in south Asia and sub-Saharan Africa. The biggest threats are more to the brain than to the throat. Children that have even mild-to-moderate iodine deficits in utero or early childhood often have stunted cognitive development because of thyroid insufficiency. It does not have to be this way. "Iodine deficiency is the most common cause of preventable mental impairment worldwide," writes Swiss thyroid researcher Michael Zimmerman.[18] At an estimated price of 2–3 cents per person per year, universal salt iodination and its benefits for the "global IQ" seem well worth the cost.[19]

THYROID DISORDERS AND SURGERY

Even in times and places where iodine is abundant, diseases of the thyroid are remarkably common. The American Thyroid Association estimates that one in eight Americans will have a thyroid condition in their lifetime.[20] Yet it is often under the radar—as many as 60 percent of those with thyroid disorders are unaware of their condition. However, in some cases, thyroid disease is quite high profile, appearing in both the political and entertainment sections of the news. George Bush Sr. and Barbara Bush both had hyperthyroidism; Hillary Clinton and Bernie Sanders have had hypothyroidism; Gigi Hadid has had hypothyroidism; and Oprah Winfrey has had both hypo- and hyperthyroidism. Some of these celebrities have been quite vocal about their condition as a way of educating the public about the unrecognized prevalence of thyroid disease.

Because of the prevalence of thyroid disorders, removal of the thyroid is among the most common of all neck surgeries. Loss of the thyroid gland itself is not so bad since patients can take oral thyroid hormone supplements. Calcitonin plays a minimal role in humans, so it is not necessary to replace. However, it is more difficult to compensate for the loss of the parathyroid glands. Life without parathyroid glands requires an expensive, daily regimen of parathyroid hormone injections. Given these difficulties, surgeons go to great lengths to spare the parathyroids when they are removing the thyroid. But this is difficult for several reasons. First, the position of the parathyroids can vary from person to person. Sometimes the lobes are detached from the thyroid and embedded in other throat tissues; sometimes one or more lobes are missing altogether. More important, the parathyroid glands are very small and similar in appearance to the surrounding thyroid tissue. In fact, the parathyroids are so obscure that they escaped the notice of human anatomists until 1880, and they were the last major organs of the human body to be named.

The fusion of the parathyroid and thyroid glands gives surgeons headaches, but for anatomists it is a beautiful illustration of a more general and widespread feature of neck anatomy. Many structures in the neck that seem like single anatomical units are in fact composites of tissues from different origins that secondarily combine in the developing embryo. To build a neck, a logical engineer might first lay out the necessary pieces in miniature form, all in their final orientation, and then just let them grow in place to generate the eventual adult form. But our bodies, and our necks in particular, are not built this way. Instead, they are built more like ceramic sculptures in which chunks of clay are smashed together, pushed around, pinched off, and smoothed over. Similarly, the embryo forms through a series of events in which tissues change shape, migrate, combine, detach, or self-destruct. The neck seems more like the product of an ornate story than an elegant plan.

Let's see how this story plays out in the migration and aggregation of the three hormone-producing tissues in the neck. Of these three sets of tissues, the thyroid hormone–producing portion of the thyroid is the earliest to form in the embryo. One of the first structures to develop in the embryo is the gut tube. At about twenty-five days of gestation, the thyroid gland originates as a pocket that protrudes from the floor of the tube near the mouth end, just as occurs in our protochordate relatives like lancelets. After pinching off its connection to the gut at the top, the gland descends as a detached mass, eventually resting on the surface of the larynx.

The parathyroid originates in an entirely distinct set of embryonic structures. At about the same time the thyroid is forming in the floor of the gut tube, the sides of the tube are folding into a series of five so-called pharyngeal arches. In fish these arches form the gills, but in terrestrial vertebrates they are redeployed to form many tissues in the neck, including the tongue, many throat muscles, the epiglottis, the hyoid, and the cartilages of the larynx. The parathyroid gland forms as two pairs of separate lobes, one pair from the third pharyngeal arch and one pair from the fourth arch.

Like the thyroid, these two pairs of parathyroid rudiments migrate downward during embryonic development. Curiously, each pair hitches rides with distinct embryonic locomotives. The top two lobes—the "dots" on the upper wings of the thyroid—travel along with the descending thyroid. However, the lower parathyroid lobes are dragged downward by yet another migrating gland, the thymus. The thymus originates in the pharyngeal arches near the rudiments of the lower parathyroid lobes and descends to its eventual position near the heart. Halfway down the journey, the parathyroid tissue unhitches from the thymus and finally parks on the lower wings of the thyroid. However, this downward journey is not always consistent. The parathyroids sometimes disembark at unusual locations, leaving them scattered at various points in the throat, much to the dismay of many surgeons.

The calcitonin-producing C-cells derive from yet a third kind of embryonic tissue and developmental process. They arise from so-called neural crest cells, which, as the name implies, originate at the top of the developing spinal cord. From there, a cluster of cells migrates downward and forward into the last pharyngeal arch, where they form structures, the aptly named ultimobranchial ("last gill") bodies, which exist only transiently in human embryos (but persist into adulthood in fishes). Later, cells of the ultimobranchial bodies fuse with the thyroid and subsequently disperse throughout its tissue.[21] For doctors trying to treat a disorder of just a single hormone, this fusion of three endocrine tissues into a complex lump is quite vexing. It's easy to imagine them cursing the anatomy. "It shouldn't be this difficult!" But the composite nature of this three-part endocrine gland perfectly illustrates the construction-by-fusion that is so prevalent in the renovated design of the neck.

• • •

A 1936 photo in the State Library of Ohio shows six people in full surgical scrubs and masks holding surgical instruments.[22] One of the doctors is George Washington Crile Sr., and the photo commemorates his twenty-five-thousandth operation removing goiter. Crile (1864–1943) was a giant in the history of surgery. In addition to his prodigious work on the thyroid, he performed one of the first surgical removals of the larynx and pioneered the removal of lymph nodes and other neck tissues in cancer treatment. But more than these specific techniques focused on the neck, he was among the first physicians to use experimental physiology to inform and improve surgical practice, earning him the title of "the father of physiologic surgery." Crile combined his experiences with patient trauma and his experiments on animals to develop techniques to minimize surgical shock, a condition when blood pressure falls as a response to the trauma and anesthesia of surgery. He came to see that patients' emotional responses to the prospect of surgery, the anxious anticipation alone, could alter their physiology and subsequently compromise the outcome of surgery. To minimize the risk of shock, surgery should be as calm, painless, and quick as possible.

Crile applied his understanding of shock to many types of surgery, but he became especially focused on the thyroid. Working in Cleveland, he was in the heart of the US goiter belt. He was repeatedly presented with cases in which the swelling of a goitrous thyroid gland became so severe that it hindered breathing and swallowing. For one of his early thyroid patients, the surgery itself went fine but the patient died from postoperative complications. Crile concluded that the patient's preoperative anxiety contributed to the development of these complications. In subsequent years, Crile developed a technique that came to be known as the "steal the gland" technique. Before the surgery the nurses daily administered small doses of anesthesia to acclimate the patient to the process of inhaling the anesthetic. Then, without forewarning on the day of surgery, the patient received a full dose of anesthesia, and Crile entered the operating room, made a small incision above the thyroid, and rapidly excised it. The procedure was over within fifteen minutes, before the patient even knew what happened. While the "steal the gland" technique was unusual at the time and would never be permitted today, it was hard to argue with its success. With this procedure mortality from thyroidectomies at Crile's hospital fell from 16 to 2 percent. Practice must have helped as well; Crile routinely performed twenty to thirty thyroid surgeries per day.[23]

THYROID HORMONE AND THE REGULATION OF STABILITY AND CHANGE

In adult humans the primary function of thyroid hormones is to maintain our thermal and energy balance. It is all about stability. Humans, as well as other mammals and birds, have the capacity to generate a lot of metabolic heat that keeps our bodies at warm and constant temperatures. Such warm-bloodedness—termed "endothermy" by biologists—is an enormous determinant of our mammalian lifestyle. It permits a high level of activity, both physical and cognitive, but it also requires that we feed ourselves nearly constantly to fuel our internal furnace. Despite living in climate-controlled buildings, humans experience daily and seasonal changes in thyroid activity to stabilize our internal thermal environment. The brain senses the thermal and energetic state of the body and adjusts the secretion of thyroid hormone to maintain long-term stability of our body temperature. The capacity of thyroid hormone to raise and regulate body temperature was surely one of the crucial adaptations leading to endothermy.[24]

If you look into our embryonic and evolutionary past, you see that the function of thyroid hormone is far more than just maintaining stability. It is largely about regulating major life transitions. In the first half of embryonic life, thyroid hormone supplied through the mother's blood promotes the production of cells in the brain, heart, lungs, and muscles. In the second half of gestation, the fetus secretes its own thyroid hormones, which coordinate the later functional differentiation of these organs as well as the liver, bones, and kidneys. These developmental actions of thyroid hormone are widespread and transformational across the body.

An even greater transformational role of thyroid hormone is seen in other vertebrates that undergo especially drastic morphological changes over their lifetime.[25] Indeed, thyroid hormones are literally metamorphic in these animals. Many frog species during their life cycle transition from aquatic, swimming, gill-breathing, herbivorous tadpoles to semiterrestrial, hopping, lung-breathing carnivorous adults, all within a matter of days to weeks. This remarkable transformation entails both the formation of new structures (e.g., limbs, lungs, and thick skin) and the destruction of tadpole structures (tails, gills, thin skin, and long intestine). Thyroid hormones drive both of these seemingly opposite processes in amphibian metamorphosis.

Because thyroid hormones have such dramatic developmental roles in humans and other vertebrates, the long list of chemical pollutants that can modify thyroid function is raising growing alarm.[26] Industrial chemicals used as flame retardants, pesticides, or plasticizers disrupt thyroid function at various levels, ranging from regulation of the thyroid hormone secretion by the brain and pituitary to its production in the thyroid gland to its binding at the tissues. One thyroid-disrupting chemical that has recently been in the spotlight is perchlorate, an additive to rocket fuel and fireworks. Perchlorate inhibits the uptake of iodine from the blood into the thyroid gland, and high doses of perchlorate can thus cause hypothyroidism in adults.[27] More concerning is that low-level perchlorate exposure during pregnancy disrupts the transport of iodine across the placenta and, by suppressing thyroid function in the fetus, it can cause neurodevelopmental disorders, with decrements in IQ.[28]

In 2011 the Obama administration enacted regulations on the concentration of perchlorates in drinking water.[29] Not even a decade later, in 2020, the Trump administration overturned these regulations. Much to the dismay of environmental groups, the Biden administration upheld this policy. Legal suits are pending.[30] Ironically, after two centuries of efforts to assure that iodine is adequate in the diets of everyone, the next public health battle may focus on eliminating pollutants that inhibit its use by the body.

• • •

Our bodies are made from a handful of earth's abundant elements; carbon, oxygen, hydrogen, nitrogen, and calcium make up more than 99 percent of the atoms in the body. These—along with other crucial elements such as sodium, chloride, and potassium—are readily available in the air, water, and food we consume. But the thyroid gland binds our fate to a single, rare element—iodine. Without iodine, the thyroid swells, and our metabolism falls. When it is missing or its uptake is blocked by pollutants during gestation, our cognitive capacities suffer. When it is radioactive, we gravely risk cancer. But when it is present, even in tiny quantities, it allows us to produce small, simple molecular signals—thyroid hormones—that play a huge and complex role in our life. Thyroid hormones act throughout the body to help keep us warm and energized, enabling our mammalian lifestyle. They help orchestrate the remarkable life transition from an embryonic blob to an ornately differentiated child with astounding cognitive potential. And in the diverse metamorphic animals around us, they coordinate even more fantastic transformations. While reliant on a rare element, the thyroid helps drive the daily pace and trajectory of life.

Nestled beneath the thyroid (sometimes inconveniently so), the parathyroid gland makes its own signaling molecule (parathyroid hormone) with a more targeted role. In our fish ancestors it likely acted on the gills to control the uptake of external calcium. In us and our terrestrial vertebrate cousins, it helps shift calcium internally between the bones and the blood and thereby helps maintain our structural integrity and temper our excitability.

CHAPTER SIX

Word & Flesh

Speech and Song at the Neck

I was impressed at how respectful everyone was of the corpses. At the end of the yearlong human anatomy class, the medical students and faculty held a nonsectarian ceremony honoring the people who had donated their bodies for dissection by aspiring physicians. I visited the human dissection lab a few weeks before the ceremony, after the cadavers had been worked on for seven months. By now, many of the tissues had partially dried out and become dark in color. I had dissected many animals, but this was my first time to see inside a human. With only the thorax and abdomen visible—the cadaver's face and lower half were covered with gauze and cloth—I could mostly separate myself from the humanness that once occupied the body and to simply explore the intricacies of the structures.

Later, my host took me to another room to look at a fresher cadaver that had been dissected with great skill by the professor to demonstrate more hidden or unusual structures. At one point the professor drew attention to a particular nerve, the recurrent laryngeal nerve, which controls the larynx (the voice box). He opened up the larynx to expose the short and very thin vocal folds. Until this moment the cadaver was simply a glorious display of anatomy. But when I saw these small cords, I was overwhelmed by the sense that this body had been a real person, with a complicated life, with thoughts and emotions that projected into the world as sound. Every utterance that ever emanated from this man over his roughly seventy years—from the most mundane ("good morning") to the most momentous ("I do")—passed through and was made possible by that small mass of tissue in his neck. Every hello, every cheer, every scolding, every lullaby, every goodbye—all of them originated there, from those particular cords. His words became flesh. Or better, his thoughts became words through flesh.

Generating vocal fold vibration is only the first step in forming words and song, but none of the rest of the steps would much matter if breath were not first converted to rapid oscillations. And, at some basic level, so much of how the mind projects itself to others is in the details of these vibrations originating in the neck.

• • •

Spoken language requires phenomenal brain power—learning, memory, syntax, semantics—but equally, it requires a vocal tract that can translate all those thoughts into a commensurate array of sounds. The position of the vocal folds low in the throat combined with an agile tongue and lips gives humans the ability to generate sounds that differ in a broad range of pitches, intensities, and timbres, all of which are rapidly modulated to create acoustic units (phonemes) with inexhaustible combinations. Even beyond speech, humans can produce a variety of other communication sounds: songs, whistles, screams, and cries. Yet, although human acoustic production is unusually rich, there are animals that exceed our vocal capacity in almost every dimension. Certain animals produce higher or lower pitches, louder calls, faster songs, raspier roars. In some cases, such remarkable vocal abilities depend on sophisticated neural control systems, but these superhuman acoustic abilities arise largely through specializations in the vocal tract, through elaborations of cartilage, muscle, and skin. The orchestral varieties of animal vocalizations—the croaks, chirps, and growls—all originate as breaths that are then translated by a vast diversity of vocal instruments.

HUMAN SPEECH: THE MECHANICS UNDER THE MEANING

Five hundred years ago, Leonardo da Vinci likened the human voice to a musical instrument, and the analogy is still useful.[1] When you look at a musical score, you see instructions for many separable features of music. Pitch is indicated by the vertical position of the note on the staff, volume by notations such as *pianissimo* through *fortissimo*, and duration by the note's time value (eighth note, half note, and so forth). All these features of sound can be produced by whistling lips and simple pipe flutes. Such "instruments" emit relatively pure sounds composed of one main tone (the dominant frequency) with overtones (harmonics) superimposed at higher frequencies. The volume of these overtones diminishes in a mathematically simple, regular way with each successive octave. A flute playing a middle-C note creates strong vibrations 256 times per second (the unit for frequency is Hertz [Hz]), with weaker vibrations at twice the dominant frequency (512 Hz), and even weaker vibrations at successive doublings, 1024, 2048 Hz, and so on. Such clean math is restricted to the small set of sounds generated by instruments with relatively simple shapes. The vast majority of vocal sounds are far messier. They are products of an instrument—the vocal tract—with many vibrational elements that quiver with varying magnitudes and that modify the strength of the overtones.

The voice, like all instruments, must have a way to generate vibrations. To make much of a sound, you need a buzz. The violinist's moving bow makes the violin's strings buzz, the saxophonist's breath flowing over the reed makes a saxophone buzz, and the larynx's vocal folds provide the buzz for the voice. When air blows on the vocal folds, it forces them apart momentarily, but after stretching open, they elastically recoil shut. They are then forced open again, over and over.

In the human larynx this happens about one hundred to two hundred times per second. When singing a middle-C note, it is 256 Hz.

In addition to their mechanisms for generating vibrations, most musical instruments, including the voice, have resonating chambers that modify the sound: the curved body of the violin, the folded and tapered tube of the sax, and for the voice, the complex shapes of the throat, mouth, nasal cavities, and sinuses. These chambers resonate each in their own way, and in doing so, they filter the sound. Each instrument has its unique way of selectively enhancing or dampening different sets of overtones, giving the sound its "tone color" or timbre. Timbre explains why two instruments, say a violin and a sax, can play the same note at the same volume but still sound utterly different. Timbre is also why you can immediately identify a family member on the phone from a single word. Because the shape and size of throats (and mouths and tongues and teeth) differ among people, even if subtly, everyone has their signature timbre, their vocal identity. Computers and phones are increasingly able to identify people by their unique voices, their voiceprints, so much so that voices will likely replace keys and passwords as guardians of our personal property and information.

With every note in a song or prosodic excursion in speech, we alter the pitch of the voice by muscularly changing the tension and position of the vocal folds. One set of muscles (the cricothyroids) pulls on a set of cartilages that stretches the vocal folds. Just like tightening strings on a guitar, this raises the pitch of the voice. Another set of muscles connects the larynx to bones above (the hyoid) and below (the sternum). They move the whole larynx vertically within the throat. This movement effectively changes the length of the resonating chamber. When the larynx moves downward, it lowers the pitch of the voice, like a trombonist extending the slide. You can feel this movement by placing your fingers on the Adam's apple (the thyroid cartilage of the larynx) while singing up and down the musical scale with your voice.

A third set of laryngeal muscles moves the vocal folds in and out of the air flow. To inhale, you open the folds to allow for unimpeded flow of the

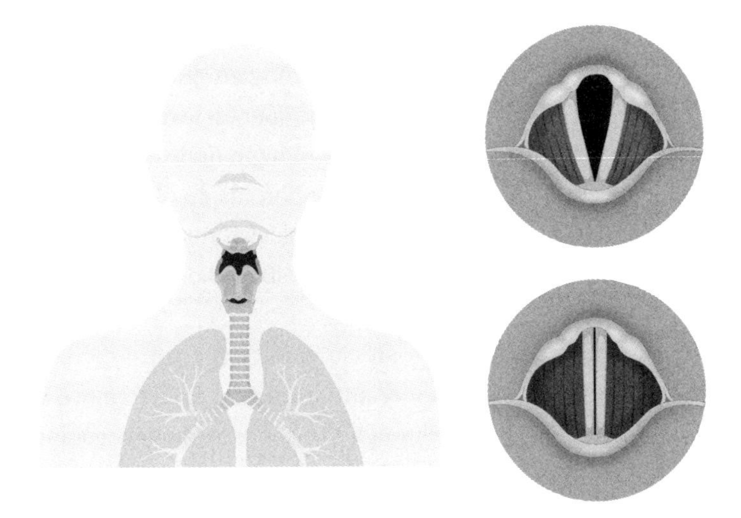

Figure 11. The position of the larynx (voice box) in the human throat. The vocal cords (folds) swing open during inhalation to allow air to pass into the lungs and close together during exhalation to produce the vibrations of the voice. Illustration from iStock.

breath into the lungs. To speak, you swing the vocal folds into the exhaled breath to generate vibrations. This action is termed phonation. All vowels are phonated, and you can feel this by touching your larynx with your fingers while saying "ahh" and "ehh." Some consonants (e.g., "t" and "s") are unphonated; they are formed only from the lips and mouth and do not engage the vocal folds. Most unphonated consonants also have a phonated equivalent. You can feel an off-on buzz in your larynx while alternating between "p" and "b" or "f" and "v" sounds. All phonation occurs when muscles move the cartilages holding the vocal cords (arytenoid cartilages) inward, placing the vocal folds into the path of the exhalation. So with every syllable change and every breath, the vocal "gear" is disengaged and reengaged by a fine muscular clutch. Altogether, the larynx has seventeen muscles that operate the position of nine cartilages.

The loudness (amplitude) of the voice results from the pressure and timing of air forced through the vocal folds. It is regulated by both breathing muscles—the abdominal muscles below the ribs and the intercostal muscles between the ribs—pushing air outward and throat muscles acting as a valve at the top. The loudest expressions of the voice come from blurting. When the breathing muscles contract while the vocal folds are closed, a large pressure builds up. Then, when the laryngeal valve (the glottis) opens, the explosive release of air causes broad oscillations of the vocal folds, and a loud shout bursts into the world.

While shouting requires high pressure and large vocal cord vibrations, our quietest voice does not require the vocal folds to move at all. When we whisper, we pass air through the larynx without engaging the vocal folds. Like so many things, this intrigued Leonardo, who annotated his sketch of the larynx with the reminder "Write on the cause of the voice without sound as do those who whisper."[2] During whispering, the uninterrupted movement of air generates turbulences within the vocal tract air column that can be modified by the shape of the oral cavity, tongue, and lips, allowing us to form intelligible speech. However, the vocal folds are not engaged, so the sounds are very weak. Compared to normal vocalization, it takes ten times as much airflow when whispering to produce the same volume of sound. You can easily feel the increase in air flow by placing your hand in front of your mouth as you switch from normal speech (soft breath with a buzz in the throat) to a whisper (heavy breath with no buzz). Or you may have felt this hot, puffing breath on your face as a lover whispers sweet nothings to you.

Unlike most musical instruments, the vocal tract continually changes shape and resonant properties as it transmits sound. For example, with every vowel we modify the shape of the throat and position of the tongue. To say "ee," we raise the tongue to the top of the palate and move it forward toward the lips, and to say "ahh," we depress and retract the tongue. The human ability to produce a broad range of vowels relies on our ability to drastically change the shape of our two-part vocal tract, or in technical terms, our "double resonator system." One resonator, the oral/nasal cavity, is horizontally oriented while the other resonator, the throat or pharynx, is vertical. In comparison to other mammals, the human oral/nasal cavity is relatively short because our face is foreshortened. But the second cavity, the pharynx, is relatively long in humans. Due to evolutionary changes in both the face and the pharynx, these two cavities are about the same length, and this 1:1 ratio enhances the range of sounds adult humans can make.[3]

We were not always blessed with a vocal tract of such favorable proportions. As newborns, the larynx is positioned high in the throat, so there is barely any space between the larynx and the mouth. The larynx descends over the first year of life, just prior to the period of rapid language acquisition, creating a large resonant cavity above the larynx. Like the larynx of human babies, the larynx of our human ancestors (and our ape cousins) was positioned similarly high, and it descended over the long period of human evolution, likely contributing to our ability for speech. While the lengths of the oral/nasal and throat cavities are approximately equal, the diameter of the cavities varies widely. During vocalization the muscles of the throat and tongue drastically change the shape of both cavities, and the ratio of their diameters can vary from 10:1 to 1:10. For example, when we say "ee," the cross-sectional area of the horizontal oral cavity is about ten times smaller than that of the vertical pharyngeal cavity. When we say "ahh," the vertical tube is ten times smaller in diameter than the horizontal tube.

No other vocalizer in the animal world has this flexibility in their throat or this range of phonated sounds. When we coordinate these shape changes with breathing rhythms of the thorax and the agile movements of the larynx, tongue, and mouth, the words in our brain become our speech into the world. The mechanical details seem dry and cold, yet this is the physics that allows us to project our most human impulses, both sublime and mundane. With this vibrating flesh, we recite poetry, sing arias, greet coworkers, and order pizza.

• • •

One way vocalization differs from other nearby functions of the neck is that it is almost entirely voluntary. The sensors, tubes, and glands in the neck operate without thought. But with rare exceptions—for example, a startled yelp—the voice relies on conscious, intentional input from the brain. Because the voice is controlled by many muscles along the vocal tract, every act of speech involves the highly coordinated action of numerous nerves. Six of the twelve prominent cranial nerves exiting directly from the brain send branches to tissues of the mouth, tongue, throat, and larynx to control vocalization. One of these nerve branches—the recurrent laryngeal nerve, which was dissected so skillfully in the anatomy lab I visited—is particularly vital for the voice. It controls all but one of the intrinsic laryngeal muscles and thereby directs the vocal folds to open, close, and change tension. Damage to other nerves can degrade vocalization, but damage to the recurrent laryngeal nerve abolishes it all together. Along with its importance for vocalization, the recurrent laryngeal nerve was featured in one of the most important experiments in the history of Western medicine. Moreover, its anatomy is so bizarre that any discussion of the neck should offer an account.

In the late second century CE, Galen of Pergamon, physician to the Roman emperor, entered a public hall surrounded by many of Rome's most prominent scholars and politicians. An assistant brought in a pig that was strapped to a table, lying on its back while struggling and squealing. In front of an audience, Galen cut into the neck of the pig and snipped two nerves running down the side of its trachea. The pig continued to struggle, but immediately after the nerves were severed, the squealing silenced. Galen had cut the recurrent laryngeal nerves, demonstrating vividly that these nerves specifically controlled vocalization. This experiment demonstrated a concept of such significance that it was recounted in medical textbooks and curricula for at least the next sixteen centuries. Before Galen, others had posited that the brain and its nerves control the body based solely on anatomical inference and clinical observations. Galen's demonstration on the squealing pig

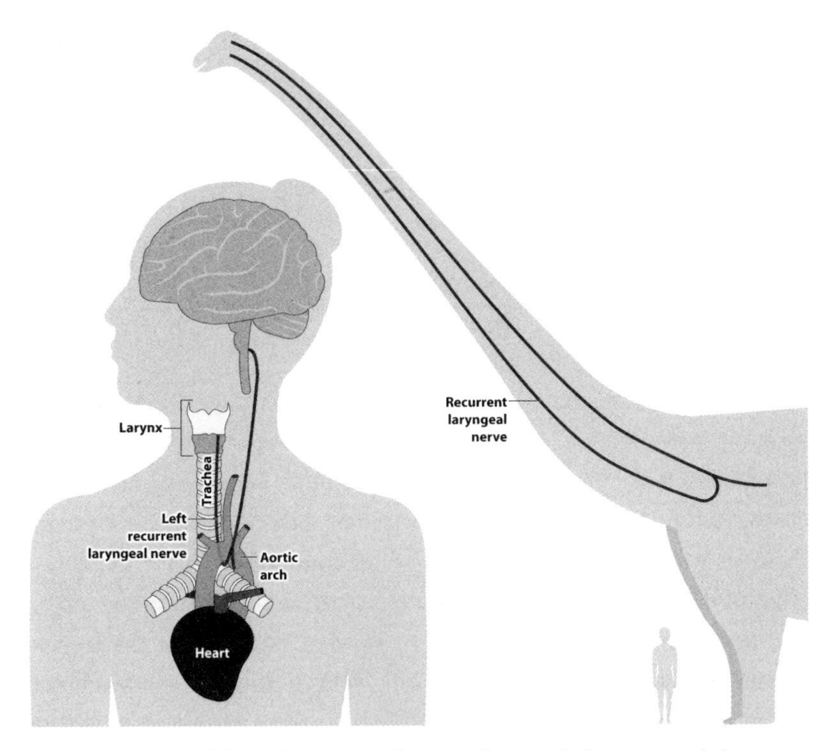

Figure 12. Path of the left recurrent laryngeal nerve in humans and the hypothesized path of the nerve in sauropod dinosaurs. The nerve descends from the brainstem, loops under the aorta, and ascends to the larynx. The neurons in the recurrent laryngeal nerve of sauropods were likely among the longest cells in evolutionary history. Illustration by Netta Kasher, 2024, modified from figure by Matthew Wedel (Wedel 2011).

was the first to prove experimentally the crucial role of the nervous system in driving behavior.[4]

Before this momentous public demonstration, Galen had dissected many necks of diverse mammals and birds in his private quarters. He found that the recurrent laryngeal nerves take a circuitous path exiting from the base of the brain, proceeding alongside the trachea, and bypassing the larynx on their way into the chest. They then loop around large arteries near the heart (the aorta on the left and the subclavian on the right), take a U-turn, and head upward back to their destination at the larynx. The nerves emerge from the skull only about 15 centimeters (6 inches) from the larynx, but rather than taking this direct route, they travel paths of about 130 centimeters (over 4 feet) on the left side or about 60 centimeters (2 feet) on the left in their looping path through the chest.[5] The neurons in this nerve are among the longest in the human body. The neurons in the recurrent laryngeal nerve of a long-necked dinosaur, measuring about 28 meters (92 feet), were likely among the longest neurons to ever exist.[6]

To us, the roundabout path of the recurrent laryngeal nerve makes no sense, but to Galen, the looping path fit perfectly with his notion of how nerves control movement.[7] He believed that nerves activated muscles by pulling on them mechanically (rather than electrically activating them, as we now know) and that muscles contracted toward the entry point of the nerve. A nerve passing directly from the brain to the top of the larynx could only pull the larynx upward. So, to move the larynx downward, as commonly occurs during speech, the nerves must enter the muscle from below. To get there, they take a U-turn around the arteries to exert a downward pull. Galen believed this looping path allows the nerves to capitalize on the mechanical advantage of a pulley.

This mechanical explanation does not conform with our current understanding of the electrical nature of nerve-muscle interaction. Instead, we find an explanation in the embryonic formation of the nerve. The path of the nerve makes no sense in adults, but it does in the embryo. At about one month of gestation, the developing head and brain are hunched over the thorax and heart. The pharyngeal arches lie between the brain and the heart and give rise to both the larynx and the great vessels exiting the heart. At this stage, when the recurrent laryngeal nerves make their connections, the brain and the larynx are quite close and, to make the short connection, the nerves happen to pass under the emerging great vessels of the heart. As development proceeds, the head starts to uncurl upward, the neck elongates, and the heart descends, dragging the loops of the nerves along with them.

The peculiar layout of this nerve is so bizarre that it is commonly cited as evidence against the "intelligent design" argument of creationists. Ironically, one of the pinnacles of human intelligence—language and speech—is accomplished through a nerve that no one could ever call intelligent in design. In our distant vertebrate ancestors the larynx arose seemingly haphazardly at a midpoint of embryonic development, after the brain and the heart had already formed in the same area. Splicing in the larynx at this stage meant its connection to the brain became entangled in the movement of the preexisting organs and tubes. These nerves to the larynx retained their idiosyncratic path throughout the evolution of terrestrial vertebrates, and in our human ancestors these peculiar nerves were the ones that transmitted the first inkling of language from the mind to the voice.

EVOLUTION OF VOCALIZATION: FROM BREATH TO SONG

The birthdate of speech is a subject of great debate. Based on different kinds of evidence, scientists have argued that speech arose anywhere from two hundred thousand years ago, with the appearance of modern *Homo sapiens*, to twenty-seven million years ago, in the common ancestor of humans and Old World monkeys.[8] However, most scientists agree that the initial steps toward any form of vocalization occurred in early aquatic vertebrates that first evolved structures for breathing air about three hundred million years ago. Fishes that respire only through gills have a rich diversity of structures for making underwater sounds, including strumming pectoral fins, gnashing teeth, and quivering swim bladders, but they do not generate sound by pushing air over tissues to cause their vibration. Only air-breathing vertebrates have a larynx, enabling them to transform air into voice.

Nevertheless, the larynx originated in rudimentary form in a certain group of fish, the lungfishes, where it regulated the flow of air for breathing. These fish supplemented the intake of oxygen through the gills by gulping surface air into the mouth, pressing it through into a short tube (the primordial trachea) and into an air-filled vascularized cavity (the primordial lungs). At the junction of the oral cavity and the trachea, these fish evolved a muscular valve to prevent the flow of water and food into the lungs. In modern lungfish some species have muscles that only shut the valve while others can actively open or close the valve, suggesting a sequence of transitions that led to increasing muscular control of airflow to the lungs. Since this earliest evolutionary origin in aquatic vertebrates, all subsequent terrestrial vertebrates had a larynx positioned atop the trachea.

While many creatures of the ancient seas no doubt communicated through sound, the invasion of terrestrial environments brought with it new possibilities for acoustic communication. It was not so much the medium of air that enabled the flourishing of sound. In fact, sound travels farther and faster in water than in air. The key innovation came as sound production became coupled to respiration and the movement of air within the body. When vertebrates began pulling air into their lungs to acquire oxygen, they could exhale the air through a complex set of cavities whose resonant properties can be quickly and intricately controlled by muscles of the throat, tongue, and mouth.[9]

As vertebrates colonized land, many of the structures formerly used for breathing in water were freed up to form structures for vocalizing. For example, the embryonic pharyngeal arches, which give rise to the gills in fish, form the larynx (pharyngeal arches 4 and 6), the throat muscles (arch 4), the lip musculature (arch 2), and the tongue (arches 1, 2, and 3) in mammals. Beginning with the earliest experiments of vertebrate life on land, evolution redeployed these throat tissues into structures of the vocal tract that transform the rather dull oscillations of breathing into gloriously elaborate and diverse vocal sounds. With a few exceptions all vertebrate lineages that prominently use acoustic communication—frogs, birds, and mammals—generate their calls through their vocal folds within the larynx and the resonant cavities above them. Peeping, quacking, barking, roaring, and bugling all are sounded from the neck.

While all terrestrial vertebrates share the original function of a larynx as a respiratory valve, the secondary conversion of the larynx into a vocal organ likely occurred many separate times in the evolution of terrestrial vertebrates. That is, the earliest frogs, birds, and mammals probably did not vocalize, and the ability to vocalize evolved independently in each lineage early in its history, about one hundred million to two hundred million years ago. A recent survey of almost eighteen hundred vertebrate species showed that these multiple origins of vocal communication often coincided with shifts to nocturnal lifestyles.[10] Sound was an auspicious mode for communication in early terrestrial vertebrates because their nighttime activity prevented them from using visual information to communicate. Later, animals in several lineages, particularly birds, began to vocalize in the daytime as well. Vocal communication was evidently quite successful; it is estimated that presently almost 70 percent of terrestrial vertebrate species call or sing from the throat.

For frogs and toads, vocal communication and respiration became linked to prominent structures just above the larynx, the vocal sacs. Frogs, unlike mammals, have no diaphragm and only very small ribs, so they cannot use these structures to draw air into the lungs. Instead, they use throat muscles to expand their mouths, pulling in air through their nostrils and pumping it down into their lungs. After air fills the lungs, it can be passed up through the larynx during exhalation to generate sound. With the mouth and nose closed, the exhaled air fills the throat and causes the vocal sac, a highly flexible structure, to balloon out below the mouth. The elasticity of the vocal sac pushes the air back into the lungs, and the air can be reused for another burst of calls. This air-recycling system has several physiological advantages. It allows frogs to capitalize on the energetic savings of using passive recoil of the vocal sac for driving air (rather than an active muscular pump) and minimizes water loss by retaining moist air for several rounds of sound production. Acoustically, the inflated vocal sac acts as a radiator to broadcast the call and as a filter to modify it. Finally, in some species the puffed-up throat serves as a prominent quaking visual signal to augment the acoustic signal during courtship, as elaborated in chapter 7.

The apogee of animal vocal performance occurs in birds. Humans have been enamored with avian virtuosity throughout history. Many familiar pieces of European classical music were inspired by bird songs.[11] In the second movement of Beethoven's Sixth Symphony recreating a scene by a brook, the flute trills like a nightingale, the clarinet coos like a cuckoo, and the oboe mimics a quail. Ravel's daybreak movement of the ballet *Daphnis and Chloe* reenacts the morning bird chorus in all its splendor and diversity. In the tradition of Iranian classical music, the nightingale (*bolbol*) reputedly has "the most beautiful voice in all of creation."[12] Moreover, "it is said to never repeat itself in its song." The nightingale is endlessly inventive and thereby embodies the Iranian paragon of *khalaqiat*, or "creative improvisation." The student must spend years mastering the highly structured canon of music (*radif*), but after a certain skill level is attained, the best musicians are prepared to mimic the spontaneous creativity of the nightingale.

As a group, birds have evolved vocal capacities that far exceed those of most amphibians and mammals. Evolution in the vocal control centers of the bird brain were crucial for giving them the ability to produce their rapid trills and warbles, to generate complex patterns of syllables, and to learn and imitate new songs. But these vocal abilities rest on the evolution of a novel vocal organ (the syrinx) deep in their throat.[13] Like other vertebrates, birds have a larynx positioned on top of the trachea that serves as a respiratory valve. But about sixty-five million years ago, birds evolved the syrinx at the lower end of the trachea, right where it bifurcates toward the lungs. The syrinx took over the role of vocalization. It evolved only once, and all birds have it. While the syrinx has many of the same basic elements of the larynx, it is an utterly new structure, not simply a larynx transposed to another position. No other group of vertebrates has a syrinxlike organ located deep in the airway that serves any alternative function, so the syrinx apparently originated for sound production from the very beginning. This de novo role of the syrinx in vocalization may very well explain why birdsong, compared to vocalization in other vertebrates, is so incredibly diverse and ornate. Unlike the larynx, which had to accommodate its simultaneous role as a valve separating the air and food pathways, the syrinx did not need to serve other functions and could evolve uncompromised by such trade-offs. In birds, evolution started with a simple branching tube, transformed it into a sonic instrument in a very few steps, and then diversified it into a remarkable orchestra of instruments. The earth has been blessed with syringeal melodies ever since.

The novelty and ubiquity of the syrinx among birds have prompted scientists to ask what birds gained by using the syrinx rather than the more ancestral larynx for vocal communication. Tobias Riede and colleagues examined airflow within the trachea and found that the region of bifurcation near the lungs, where the syrinx is positioned, is characterized by especially high and complex forces on the tracheal walls.[14] Thus placing the vocal organ at this low junction may cause a greater range of vibrations in the nearby vocal structures and thereby a greater array of sounds. Moreover, when researchers piped air through artificial models that mimicked vocal organs, they found a large efficiency advantage in placing the sound-producing tissues low in the trachea in the syringeal position, as compared to the higher position of the larynx. That is, the syrinx gives more sound per breath. Such efficiency is likely very important for a small creature that seeks to communicate over a long distance.

While these airflow considerations may explain the advantage of the low position of the syrinx, they do not account for why only birds capitalize on these same advantages. The answer may well lie in the great elongation of the avian neck. (As discussed earlier in this book, birds evolved a long neck in conjunction with their bipedal stance, their relatively light heads, and their specialized respiratory system.) Their elongated neck provides a long passage for air above the syrinx. Based simply on physics, a long pipe extending from a sound source is more efficient for transmitting sound than a shorter pipe. With their relatively short necks, mammals could never achieve this efficiency boost, even if the vocal organ were positioned lower in the trachea. For example, moving the human vocal organ 12 centimeters (5 inches) down from the current laryngeal position to the bottom of the trachea (the syringeal position) would not confer much benefit. But for birds with their especially long necks, such repositioning of the vocal organ would place it in the sweet spot for capitalizing on these efficiency advantages.

In order to fly, most birds are small, which limits certain features of their voices—for example, their deepest pitch. However, their small body size also contributes to one of their most outstanding vocal skills: speed and agility. Many small birds, especially songbirds, sing remarkably fast with a wide array of syllables, each of which can have rapid changes in pitch (e.g., warbles and trills) and timbre (e.g., whistles and squawks). In addition, they rapidly string together different syllables into complex songs. To the human ear, these songs are so fast that it is often difficult to appreciate their intricacy. But if you replay recordings of these birds at reduced speed, you can hear that every second contains complex contours of pitch and timbre, melody and harmony.

In many cases, such complexity and precision carry important meanings. In some species, females choose their mates based on the temporal details of these songs. Subtle acoustic variations can mean the difference between friend and foe. Songbirds, because of their small body size, have a small syrinx that can operate at high speed because each change of tone or timbre requires only a miniscule muscular adjustment, and they have evolved some of the fastest contracting muscles in the animal world.[15] The high-frequency sounds produced by small birds can be rapidly modulated such that one syllable does not interfere with the next syllable. By analogy, the most rapid complex melodies in an orchestra are produced by the small, high-pitched flutes and violins, not the large low-pitched tubas and string basses. Small birds may not be able to roar, but they can condense a remarkable amount of sonic change, and thus information, into each moment of song.

In mammals, like in frogs, vocalization begins with vibrations in the larynx. But, compared to amphibians, mammals have a much more complicated and physically malleable vocal tract above the larynx that can filter and resonate with the vibrations. Much of the complexity of the mammalian vocal tract evolved along with a uniquely mammalian behavior, breastfeeding. To suckle, mammals evolved mobile lips, a fleshy, agile tongue, and highly maneuverable throat muscles. When not involved in suckling, these structures can be contorted into various shapes to alter the acoustic properties of the vocal tract. In addition, as part of their high-energy, high-temperature lifestyle, mammals evolved ornate nasal passages that moisten and warm the large volume of exterior air before it passes into the lungs. These nasal cavities became important resonating spaces that modify sound. Simply pinch your nose closed while you talk to get a sense of how the nasal passages influence the sound of the voice. Although nonhuman mammals cannot match birds in their vocal repertoire, the structural elaborations and high level of muscular control along their vocal tract certainly give them a more elaborate set of vocalizations than their amphibian and reptilian counterparts.

Within the primate lineage of mammals, many species have air sacs in their vocal tract that in some ways harken back to those of frogs and toads. Just under their chin they have air-filled protrusions that can filter or radiate the vocal output.[16] The amplifying ability of such air sacs is familiar to many inhabitants of New World tropical forests. The piercing howls of howler monkeys radiating from bulges in their throats are broadcast widely from the treetops and heard across long distances in the forest; howler monkeys are among the loudest of all animals.[17] While the full effect of air sacs is not well understood, it is likely that, although they amplify the voice, they also blur acoustic features of certain vowel sounds, thereby limiting the number of distinct phonic units.[18] The great apes (e.g., chimpanzees and gorillas) all have air sacs, with one notable exception: us. Our human ancestors lost their air sacs, and this loss may have permitted the expansion of the human vocal repertoire.

A second unique feature of the human vocal tract is found in the larynx itself. This feature also emerged as an evolutionary loss. In a recent survey of more than forty primate species, all except humans have thin vocal membranes connected to the vocal folds.[19] These membranes appear to enhance their capacity to generate loud, high-pitched vocalizations, but they also destabilize the vocal output. The absence of these membranes in humans allows us to avoid such acoustic chaos and to generate the stable, controllable sounds used in speech. Thus it appears that the evolution of agile vocalizations in humans was facilitated by the evolutionary loss of both air sacs and vocal membranes. That is, the complexity of human speech followed from anatomical simplifications in the throat.

A third distinctive feature of the human vocal tract traditionally associated with speech is the descended larynx. For decades it was commonly thought that the evolutionary descent of the human larynx was a springboard to human speech because it created the "double resonator" configuration that is so versatile for producing the distinct speech sounds. Recently, however, several lines of evidence have called this idea into question.[20] First, laryngeal descent is not uniquely human. Researchers in recent decades have found that several species of deer and cats as well as the koala have a descended larynx, and in none of these cases does their unusual laryngeal position confer the ability to produce speechlike sounds. Second, researchers have reexamined the dynamics of vocal tract shapes in monkeys to assess the range of sounds that an animal could make even without a descended larynx.[21] From X-ray videos of a macaque monkey engaged in vocalizing as well as chewing and swallowing, researchers found that the range of vocal tract shapes in naturally behaving monkeys was broad enough to potentially produce nearly all the sounds present in human speech.

So perhaps the descended larynx is not so unique in its functional capacity after all. While still the subject of debate, this idea has recentered explanations for the evolution of speech to the neural control of the vocal tract, not the anatomy of the vocal tract itself. Perhaps the crucial innovations were in the language centers of the human brain. And surely one of the most important of these cognitive skills was the ability to modify vocalizations through learning.[22]

VOCAL TRAINING

By the age of three, toddlers have learned to articulate almost all speech sounds and produce two hundred to one thousand words, often stringing them into two- to three-word phrases governed by sophisticated syntactic rules. But our vocal learning is not confined to speech development in early childhood. At many life stages humans learn a wide range of vocal skills: how to pronounce new languages, imitate other people's (and animals') voices, project our voice publicly, sing, hum, and whistle. Professional vocalists (e.g., singers, actors, orators, auctioneers, and ventriloquists) spend whole lifetimes training their vocal skills. For other professions that take a high level of motor skills (e.g., athletes, dancers, musical instrumentalists), it is usually easy to observe how practicing certain motions leads to mastery in their specialized movements. But for professional vocalists their skills are mostly hidden in their throat. What exactly does it take to train the voice?

Voices are acoustic signatures, and just as there are no ideal signatures, there are no ideal voices. Nevertheless, cultures throughout the world have sought to develop the voices of their prized singers to achieve certain acoustic qualities through vocal training. In the formal European classical tradition, one aim of vocal training was originally a matter of practicality. To be audible and intelligible to a large audience, singers needed to learn how to make loud, clear sounds that spread throughout the amphitheater or auditorium. In the centuries before electronic amplification, performers had to fill large spaces using just the power of their respiratory muscles driving airborne vibrations through the resonant cavities of the chest, neck, and head. The necessity of producing voices of such power likely restricted the range of vocal styles in the past. Conversely, the diversity of vocal styles in contemporary performance artists is partially a product of modern audio technology. Some of the wispy or gravelly voices in modern pop music would not have been audible or intelligible in the performance spaces of the past. They persist by virtue of the microphone and loudspeaker.

Much of the skill in producing a strong, clear voice comes from learning how to generate and regulate the breath through the respiratory muscles of the thorax and abdomen. However, I will trace the breath beginning in the throat and continuing up to the head. For the vocalist trained in classical music, one of the primary jobs of the throat musculature is to relax and get out of the way. By relaxing, the throat opens up to create a larger resonating space, and the walls of the throat become looser and vibrate with a greater dynamic range. In this configuration there is simply less impedance in getting the breath out and into the world. Tense muscles of the throat can hinder the full movement of the larynx and thereby restrict the range of possible vocal sounds.

As the voice rises into the upper throat and mouth, it encounters an enormously influential and skilled muscle: the tongue. Indeed, the tongue is the sculptor of the voice. During singing, as in speech, the tongue morphs the vocal tract with each syllable to create vowels with different timbres and consonants with variable timing and qualities. The agile tongue has a central role in making songs, but it can also cause problems. "The tongue is like a juvenile delinquent," says Joanne Scattergood, a vocal performer and teacher of voice for more than three decades. "You have to give it something to do, or it just gets in trouble."[23] To produce the clearest vowel sounds, the tip of the tongue should be placed along the back of the lower front teeth. Keep the rest of the tongue low, forward, and out of the way. In this "yawn position," the vowels are formed more in the front of the vocal tract and can be projected more easily from the mouth. If you let the tongue slide backward, as often occurs in speech (at least in English speech), the vowel sounds get muddled. For a classical vocalist this is trouble.

Soon after the musical note is sculpted by the tongue, it is launched into the world. We commonly think of voice exiting the body through the mouth, and certainly the lips and jaw shape the final output. But an important skill developed through vocal training is to channel the voice into the additional resonant cavities of the face, the area that vocalists call "the mask." Through subtle muscular modifications of the vocal tract, the trained singer focuses vibrations through various places in the palate where the vibrations are transmitted through the skull to the sinuses in the cheek, nose, and forehead. Creating such "mask resonance" is especially important for sopranos in projecting very high-pitched notes. Sopranos-in-training are often told to visualize concentrating their breath into an imaginary column running upward just outside their face. When sopranos ascend a musical scale, they sometimes imagine pushing the sound up the column, and this visualization technique helps them send vibrations into ever higher sinuses that resonate with higher frequencies. Capitalizing on these additional resonance spaces allows sopranos to produce especially strong, high-pitched notes that otherwise would excessively strain their vocal folds.

One challenge for voice students is that they cannot see most of the anatomy they must learn to control. Like all of us, they cannot even hear what their voice sounds like to their audience. As any recording of our own voice reveals, the voice we project to the world differs substantially from the one we hear in our own heads. Without visual or reliable auditory feedback, voice students must rely on how their vocal tract feels. "The way the voice feels to the student is the way it sounds to everyone else," explains Scattergood. From her own training under the tutelage of renowned soprano Phyllis Bryn-Julson, Scattergood recounts a breakthrough in her vocal studies that corresponded to a new feeling she discovered in her throat. As a soprano seeking to project the highest, loudest notes, she learned that the most important dimension of the vocal tract was its length, not its height (i.e., the widest mouth opening). Lengthening the throat gave her larynx more room so that it could more easily stretch the vocal folds with enough tension to produce the highest pitches. With these adjustments her voice simply "felt easier"—more sound with less strain. Now, with her own students, she reminds them, "If you want to sound clear, make it feel easy."

Often, singers' prowess is measured by the precision or range of their voice: how well they reach or match a given note, or how loudly they can belt it out. But in many cases, such vocal parameters are only incidental to the emotional power of a song. In his book *This Is Your Voice*, John Colapinto describes how it is often imprecision and instability that give singing voices their affective impact.[24] He chronicles some of the first research to examine the acoustic basis for the emotional power of song. In the 1920s Carl Seashore and Milton Metfessel acoustically analyzed songs, mostly spirituals sung by African Americans in Georgia, Tennessee, and North Carolina. They cataloged certain bends in pitch that conveyed sadness and vocal qualities that suggested nearly unbearable grief. But perhaps the most distinguishing characteristic of these highly emotive songs was their vocal "imprecision."[25] That is, singers hit notes slightly "off key" or the timing was slightly advanced or delayed. Such deviations in pitch and beat were irregularities but certainly not mistakes. They gave the music its emotional impact and artistry. Seashore and Metfessel subsequently found similar musical "imperfections" in the vocal performance of highly trained classical singers. So, in both folk and classical music, much of a singer's talent comes from learning how to elicit feelings by deviating—just slightly—from the expected.

Another vocal maneuver that amplifies the emotional intensity of a song is a form of intentional vocal instability. To punctuate and elevate a long note, singers often include vibrato—a rapid, naturally occurring flutter between a high and low pitch. Among opera singers, voices that include vibrato are perceived as more emotionally expressive than those without.[26] Vibrato is not specific to one kind of emotion; it is an all-purpose intensifier of emotions ranging from melancholy to fear to exhilaration.[27] Vibrato is achieved by the singer rapidly pulsing respiratory, laryngeal, and vocal tract movements. While the tremble of vibrato is unsteady, it is hardly random. Across diverse musical genres (classical, pop, country), vibrato consistently wobbles at 5–7 Hz through a semitone of pitch undulation—the same tonal difference between a white key and an adjacent black key on the piano. Perhaps not coincidentally, such oscillations are also found in two very human spontaneous emotional expressions: laughing and crying.

HUMAN AND ANIMAL VOICES COMPARED

Through years of practice, accomplished vocalists have the power to take us to a wide range of emotional places: the depths of the blues singer, or the heights of the opera soprano. Yet, for all our vocal potential, human voices have intrinsic limitations based largely on the physics of the vocal tract. We can only go so low, high, loud, and fast. But evolution has had more than one hundred million years to diversify animal voices in seemingly infinite dimensions. Human speech may be the most versatile, information-dense acoustic communication system on earth, but the range of sounds our species can make is a small fraction of that present in the animal world. The babel of animal communication is not always easy to decipher, but its grandeur is never in question.

At the low end of the sound spectrum, the deepest sounds humans normally make during speech are about 100 Hz for males and an octave higher at about 200 Hz for females. (For reference, the lowest key on a piano is about 30 Hz.) In principle, there would be a great advantage if we could produce even lower-pitched voices. Low sounds travel through the air far more efficiently than high-pitch sounds, and if we could go even lower, our voice would travel unimpeded a far greater distance. But using our system of vocal production, the voice can only go so low because of size limitations of the vocal folds.

Tim Storms is the exception that proves the rule. Storms sings bass in performances all over the world. He has been singing in this low register since he was eight years old. As a boy listening to the church choir, he noticed that he could match the lowest of low voices. As an adult, his voice grew even deeper, and he can now produce tones that are octaves beneath the lowest note on a piano, so low they are below the range of human hearing (lower than 20 Hz). In his rendition of "Amazing Grace," for instance, it almost sounds like he is purring at certain points. After one of his performances in a Christian a cappella group, an ear-nose-throat doctor from the audience approached Storms. Amazed by the depth of his voice, the doctor said, "I've got to take a look at your vocal cords. Can you come to my office?" Storms agreed. When the doctor viewed his larynx through the scope, he found that Storms's vocal cords were almost twice as long as normal.[28] Anomalies of the vocal apparatus—even if it is only a few extra centimeters—can have thunderous consequences.

The anomaly of Tim Storms notwithstanding, the depth of vocalization is closely associated with body size across all vertebrates, including humans. (Storms is not especially large.) Large animals generally have longer vocal folds that can oscillate at lower frequency. This association is fairly intuitive: large dogs *woof* and small dogs *yip*. But a purring cat breaks this rule, making a low rumble that is several times lower in pitch than even the speech of its much larger human companion. Cats purr using a mechanism entirely different than the one that humans use to talk and that cats use to meow.[29] Instead of using exhaled air to generate vibrations in their vocal folds, purring cats rapidly contract muscles that quiver their larynx at about 25 Hz to make their low-pitched hum. The largest of all terrestrial vertebrates, elephants, have a voice so deep that scientists were uncertain how it was produced. In addition to their familiar loud, audible trumpeting, elephants commonly produce sounds in a frequency range (1–20 Hz) that humans (or almost any other animals) cannot hear. With these deep rumbles elephants can communicate long distances—up to 10 kilometers (6.2 miles)—to a lost calf or a territorial competitor.

These so-called infrasonic vocalizations were discovered by researchers in the 1980s, but for the next three decades it was not known how elephants generate the sounds. In 2012, when an elephant died from natural causes in the Berlin Zoo, researchers dissected out the larynx and in the laboratory attached it to a "pseudolung" that simulated the airflow of an elephant's natural exhalations.[30] They then filmed the vocal folds and recorded the sounds that emerged. The excised larynx produced sounds at the same low pitch found in the call of a living elephant. Thus it appears that the brain (or any other highly specialized mechanism) is not necessary to produce the elephant's uniquely deep rumble. Its unmatched ability to communicate infrasonically over such long distances is simply a matter of body size. With their huge larynges elephants have long vocal folds that can generate deep rumbling voices.

• • •

When humans communicate through speech, we spend much of our conscious attention decoding the meaning of the words while subconsciously gleaning information about the speaker from nonverbal features of the voice. For example, we may attend to the melody of their speech to infer the speaker's emotional state or the quality of their voice to estimate their age. Body size is one such attribute detectable by just listening to the voice. We are generally quite good at guessing someone's stature even if we have only spoken to them on the phone. Even people blind from birth who have never visually experienced anyone's body size can estimate a speaker's body size solely by listening to their voice. While deeper voices usually mean larger bodies, we can certainly think of individuals that do not follow this trend. Researchers in bioacoustics have found that a voice parameter slightly more complicated than pitch—the spread of frequencies—helps improve our estimations of body size.[31]

On average, small people have shorter vocal tracts, and according to basic physics, when sounds vibrate through short pipes, the various simultaneous frequencies (the timbre) spread widely apart (e.g., 200, 1000, and 2000 Hz). In long pipes and long vocal tracts, the frequencies are compressed into a smaller range (e.g., 200, 400, and 600 Hz). This frequency spread can therefore be used to estimate body size: the smaller the frequency spread, the larger the apparent body size. People seem to know subconsciously that frequency spread as well as absolute pitch are important correlates of body size. When asked to imitate the voice of a larger person, vocalists decrease the frequency spread as well as lower the pitch of their voice. However, listeners are also quite good at detecting imitators. That is, when we try to fake it, we are not good at misrepresenting our body size with our voices.[32]

Some birds, however, seem to be chronic liars; they misrepresent their body size through their voices. Rather than making behavioral imitations, they use a remarkable anatomical adaptation of their tracheas to broadcast a larger-than-life voice. The tracheas of these birds, after emerging from the lungs and syrinx, spiral into loops near their thorax before rising up the neck toward the mouth. Such tracheal elongations are not rare. Since their original description in the European crane by Emperor Friedrich II in the thirteenth century, this peculiar anatomy has been documented in more than sixty avian species in widely divergent families, and it has likely evolved many times over bird evolution.[33] The most extreme case is in the trumpet manucode, a bird-of-paradise in New Guinea. Its trachea coils five full revolutions within the body before rising through the neck. Most commonly, and especially in many ground-dwelling, chickenlike species, the coils lie just under the skin of the breast. In the more bizarre cases, such as in the large cranes and swans, the trachea penetrates and spirals around within the breastbone. The unwound length of the trachea can span up to 1.5 meters (about 5 feet) in the whooping crane, and in some species it exceeds the length of their entire body.

Because tracheal elongation is found most commonly in very boisterous species, scientists have assumed for centuries that it served to amplify and/or lower the pitch of their vocalizations and thereby increase the range over which they could be heard. And, indeed, some of these birds have a phenomenal ability to project their calls. Superfi-

Figure 13. Elongation of the trachea in four bird species: (*left to right*) crested guinea fowl, European spoonbill, trumpeter swan, and trumpet manucode. Illustration by Tecumseh Fitch (Fitch 1999).

cially, this makes sense: the loudest and deepest musical horns (e.g., tubas, baritone saxophones) have the longest tubes. Also, by passing the trachea near or into hollow cavities of the body (the breastbone or air sacs), these birds might be taking advantage of resonance chambers that amplify sound much in the same way that the hollow body of a violin or a guitar does. But in his thorough analysis of the phenomenon, Tecumseh Fitch concluded that the primary function of tracheal elongation is not about vocal depth or loudness; it's for outright deception. Birds with extralong tracheas produce calls that make them sound larger than they really are. Fitch termed this the "size exaggeration hypothesis."

Fitch showed that among bird species, there was no correlation between a species' tracheal length and the pitch of its call. Birds with these elongated tracheas do in fact often make low-pitch calls, but this can be explained because they also tend to have particularly large syringes with long syringeal folds that vibrate like the long strings of a cello. Their large syringes, rather than their long tracheas, likely account for these birds' low-pitched voices. Fitch also argued that any sounding-board effect of the hollow bones and air cavities would be completely dampened by the overlying muscle and skin tissue. For this sort of amplification through resonance chambers to work, there needs to be an unimpeded way for the sound to exit the body.

Based on acoustical theory as well as audio recordings from birds, Fitch argued that the salient feature of a long trachea was its ability to enhance the call's frequency spread, not its pitch or loudness. Species with elongated tracheas make calls with a tighter collection of frequencies. Fitch believes that birds, like humans, estimate body size based on frequency spreading (the smaller the spread, the larger the estimated size) and that certain bird species evolved longer tracheas to exaggerate their perceived size, especially in territorial competitions. In support of the idea that such exaggeration functions in territorial interactions (rather than, say, sexual attraction), Fitch notes that within a species, sex differences in tracheal elongation correspond to sex differences in territorial defense. That is, for species in which males do the guarding, they alone have elongated tracheas. In the one species in which only females guard the territory, they alone have long tracheas, and in species in which both sexes guard territories, they both have elongated tracheas. Species with elongated tracheas almost always set up their territories in visually hidden habitats—for instance, in dense rainforests or in wetlands with tall grasses. Thus tracheal elongation may have evolved as a way to deceive territorial competitors into thinking their rival is especially big when they have no way to visually confirm their actual body size. This curious adaptation, in effect, enables birds to deceptively puff up their chest when they can be heard but not seen.

• • •

To avoid competition, individuals of many species guard territories, or at least live in their own separate home ranges. Often, this means that they are spaced far apart from each other. Some animals, even ones with small body size, can project their calls to members of their own species across hundreds or even thousands of meters. They seek to attract faraway mates, repel distant competitors, or simply broadcast their presence across their domain to whomever can hear. They spend a good part of their day hollering. These species exceed humans resoundingly in the loudness of their voice.[34]

To pinpoint how birds and mammals specialize their vocal production for long-distance communication, Ingo Titze and Anil Palaparthi developed a model that incorporated both the physics of sound propagation and the detailed anatomy of the vocal tract.[35] First, they found that long-distance specialists typically generate vocalizations that focus energy at relatively high frequencies (1000–5000 Hz in animals compared to 100–300 Hz in humans). Such sounds are less absorbed by the walls of the vocal tract. Second, many animal vocalists have relatively large mouths that they open quite wide when calling. For example, the loudest bird ever recorded, a white bellbird, has a gape size when it is calling that is much larger than a speaking human, even though it is only about one-six-hundredth the body size.[36] When humans want to project their scream as far as possible, they use these same two modifications: high pitch and wide mouth. Imagine someone shrieking in terror. Many birds and mammals, however, use a third technique impossible for humans because of our constrained neck anatomy. They retract their head toward their shoulders and use their whole body as a baffle, much in the same way that the box surrounding a loudspeaker amplifies and redirects sound forward. Together these vocal strategies give many loud, long-distance specialists near 100 percent efficiency in converting the power of their exhalation into vocal signals.

In contrast to these widely dispersed species, humans and other social species spend time quite near each other and usually converse at relatively low volume. Within this close range, we might vary the loudness of our voice for emphasis ("I LOVE ice cream!") or privacy ("Psst, you have some ice cream on your cheek"). However, occasionally, we let out a shout that might transmit across the building or blare out a call to get the attention of someone far away. But casting one's voice over such long distances takes great effort. While humans are very articulate in our speech, we are not very efficient in our voice. Our vocal tract (throat, mouth, and lips) enables us to generate a wide range of complex, information-dense phonemes, and such sounds are accomplished best using the low frequencies that predominate in human speech. However, our highly articulate vocal tract also dissipates or absorbs most of the acoustic power in the voice, especially because our relatively small mouth opening reflects most of the energy back into the throat. We could improve this efficiency if we spoke at higher frequencies, but we would lose some of our ability to effectively separate consonant and vowel sounds and to encode qualitative subtleties. All this to say, the natural human voice maximizes intelligibility but limits our acoustic efficiency and ability to communicate over long distances. In stark contrast to some birds and mammals, we are less than 10 percent efficient in transforming breath into voice.

• • •

Most human parents would argue with the idea that humans cannot vocalize very loudly. They have heard the powerful voice of their babies. One of the first things babies do is announce their arrival into the world with all their neonatal might through their first cries. Most people respond to the cry of children with alarm, or at least concern. But, for Maggie Pollitt, in Tennessee Williams's play *Cat on a Hot Tin Roof*, their cries—and the necks that produce them—elicit a less charitable sentiment. "Maggie the Cat" is a married but childless woman who has nothing but contempt for her young nieces and nephews. She directs her contempt at their necks.[37]

> One of those no-neck monsters messed up my lovely lace dress so I got to cha-a-ange! I swear they've got no necks. None visible. Their fat little heads are stuck on their fat little bodies without a bit of connection. An' it's too bad, 'cause you can't wring their necks if they've got no necks to wring!

> Yep, they're monsters, all right.... Hear them screaming? I don't know where their voice boxes are located since they have no neck.

We all have at least a bit of Maggie in us. The power of a child's scream has surely driven all of us mad at some point.

Acoustically, the power of a baby's cry lies largely in its chaos.[38] Most verbal sounds are an ordered collection of simultaneous frequencies. Each of these frequencies is fairly distinct, and one of the frequencies, usually the lowest, is dominant. When these sounds are received by the listening ear, they excite correspondingly distinct regions of the hearing organ in the inner ear (the cochlea) with a high level of activity in the region activated by the dominant frequency. By contrast, the infant's cry is a blast of disordered, random frequencies, which simultaneously stimulates a smear of locations on the cochlea. Babies can produce these outbursts, in part, because of their unusual vocal apparatus. The inner strand of their vocal folds is more like a gel than a string, and the edge where the vocal folds meet is poorly developed compared to that of an adult. When the vocal folds of a baby are activated by a strong burst of breath, they tend to vibrate somewhat chaotically, generating the cry's raspy quality. In addition, the vocal folds of a baby are also much shorter than those of an adult, so overall they vibrate at higher frequencies (more than 500 Hz as opposed to 100–200 Hz). These relatively short, floppy vocal folds generate the familiar, if not disturbing, high-pitched grating sounds of a cry.

Floppy vocal folds also generate the powerful vocalizations of a rather different creature: a lion. While a helpless baby and a fierce lion differ in so many ways, they share the capacity to command a lot of attention through their throats. The vocal folds of a lion, like those of a human infant, generate attention-grabbing sounds by vibrating with a broad smear of irregular frequencies. The joining edge of their vocal folds are composed of a fatty layer, which oscillates in a disordered manner rather than the "clean" oscillations of a taut cord. The biggest differences between a baby's cry and a lion's roar are in the pitch and volume. The vocal folds of a lion are about ten times the length of those in a human baby, so lions can roar at a proportionately lower frequency (50 versus 500 Hz). The loudness of the lion roar, which is about twenty-five times the volume of a gas lawn mower, is attributable to the detailed shape of the vocal folds. Their folds meet with flat, squared edges, and biophysical studies show that with this configuration they require relatively little energy to vibrate in high-amplitude waves. That is, lions can generate a very loud roar without a lot of air pressure from their thorax. While babies can effectively draw in their parents from across the house, lions can broadcast their fierceness far across the savannah.

• • •

The singular voice of a poet or a vocal soloist can be enough to move us to tears as well as a wide range of other emotions. But humans are also deeply moved by the interplay of multiple voices. Humans across time and place have brought together vocalists to sing different but complementary songs in groups ranging from duets to concert choirs. Each singer has a single voice. This statement, which may seem like truism, has exceptions in certain human cultures. Some people can simultaneously produce two voices. Found notably in the Tuvan people of Mongolia, Buddhist monks in Tibet, Inuits in Canada, and the *canto a tenore* singers of Sardinia, such "overtone singing" or "throat singing" commonly sounds like a melodious, high-pitched tone riding on top of a low-pitched constant drone.[39] Among the Tuvan people, singers develop this skill from childhood through a long apprenticeship in which they learn how to alter the shape of their vocal tract and regulate the flow of breath.

To figure out how a single throat can produce a polyphonic voice, a team of Canadian and US researchers studied three singers of the Tuvan performance group Huun Huur Tu.[40] To visualize the vocal mechanisms, the researchers used magnetic resonance imaging (MRI) while the musicians sang in a certain style of overtone singing, Khoomei. The researchers found that the low-pitched tone was generated much like the song of a typical singer, but the high-pitch whistle was generated by a distinct mechanism. Through active modifications of the vocal tract, overtone singers can merge energy from several of the higher overtones into a single pitch. This effectively filters out the highest harmonics, focuses the acoustic energy in a narrow frequency range (1000–2000 Hz), and gives the high tone its whistle-like sound. To achieve this two-tone sound, the singers morph their vocal tract by making two separate constrictions in their vocal tract, one at the top using their tongue and another deep in their throat. Through learned contortions of their throat, these Tuvan singers and singers in a few other cultures can circumvent a near-universal limitation of one voice per person.

While a select few humans have learned to make polyphonic songs, the animal world has an array of species that routinely sing with two voices from one throat. Some can even perform internal duets. In many bird species the syrinx extends down into the two bronchi (the tubes leading to each lung). Each branch of the syrinx has independent nervous control and produces different sorts of notes. Like human duets by two people (say, an alto and a tenor), individual birds can generate internal duets composed of two voices from separate syringes with distinct but overlapping pitch ranges. The right syrinx generally sings the higher voice (the alto), and the left side sings the lower voice (the tenor).

In some species the right and left branches can sing simultaneously (polyphony) and alternate their notes or syllables sequentially (antiphony). Take, for instance, just one short segment of a brown thrasher's song composed of four syllables.[41] The first syllable starts with both left and right portions of the syrinx singing simultaneously at slightly different pitches—a brief harmony of two voices. This is followed by three separate, alternating notes: a high, upward sweep from the right syrinx; a lower downward sweep from the left; and then a flat, intermediate pitch from the right. Each of the next two syllables are composed of individual notes made sequentially by the right and then the left syrinx—a two-voice antiphony. The last note returns to a mixture of three simultaneous and alternating notes each with distinct pitch contours. All this in less than 1.5 seconds. Now imagine the thrasher singing all morning long.

Such internal duetting is more than vocal acrobatics; the details of these complex two-part harmonies can have real consequences for a bird's sex life. In canaries, females prefer to mate with males that can produce rapidly repeated syllables with multiple notes and a broad pitch range.[42] In such "sexy syllables," the notes are incredibly fast—more than fifteen notes per second—and they sweep downward almost two octaves (from about 7500 to 2000 Hz). In these few syllables alone, canaries far exceed anything that the best human vocalist can do. To generate such notes, canaries use both sides of their syrinx in a tight sequence. The right side starts with the highest pitch and sweeps down the first octave. Then, seamlessly, the left side takes over and sweeps down the remaining octave. This repeats over and over, perhaps thirty times over the next few seconds. Each side of the syrinx is controlled by its own side of the brain, so this two-part song takes a remarkable degree of bilateral coordination between the brain hemispheres. Female canaries find such vocal agility alluring. In fact, researchers have hypothesized that female canaries base their mating preferences not so much on a single component of the song (e.g., speed or pitch) but rather on the song's display of bilateral coordination. That is, it is the skill in duetting—the interplay of voices within a single throat—that most impresses.

• • •

Who knows how thoughts and emotions originate in our minds. Their genesis will probably always remain a mystery. But once they arise, we transform some of these abstractions into acoustic expressions—words and songs—that can span the gap between our minds and the minds of those around us. We send elaborate patterns of electrical activity from our brains to the dozens of muscles of our chest, throat, and face that configure the shape of our supple vocal instrument and simply blow. It's an alchemy of breath into meaning via flesh. The abstract becomes material. It is truly one of the marvels of the human neck.

Hundreds of millions of years before the first word was ever spoken, land-dwelling animals began co-opting their exhalations—respiratory by-products—into chatter and song. Acoustic signals can be projected across great expanses and through cluttered environments. They are effective by day and night and can be activated or silenced in a split second. Animals capitalize on these advantages of sound in so many diverse ways. Beginning with an organ that originally served to protect the lungs, early terrestrial vertebrates evolved the larynx into a vibratory apparatus replete with expressive possibilities. Animal species have broadened their spectral range from the infra- to the ultrasonic. They have supplemented their vocal tract with new resonating cavities or even added an entirely novel vocal organ. They have changed the dimensions and muscular control of the vocal tract to vary the voice along practically every conceivable parameter. "Songs are really just interesting things to be doing with air," said the singer Tom Waits. Nature has elaborated these "interesting things" with unending creativity using a dazzling diversity of vocal instruments. We were preceded by and cohabit the earth with a choir of humbling proportions.

CHAPTER SEVEN

Courtship & Attraction

Sexual Communication at the Neck

A mute swan flies toward a lake, its long neck stretched out straight, counterswaying with the undulations of its body to keep its head stable. After landing, the swan turns its head clear around to preen its wings then dips its head deep into the water to feed on aquatic plants on the lake floor. It paddles across the water, head held high to scan the surroundings, waddles onto the shore, and twists its neck around to rest its head on its back. In this short sequence in the solitary life of a swan, the neck has operated with such versatility, helping it to sense, groom, feed, and rest. But this particular swan has come here to do something necessarily social, to mate, and its neck—and the necks of other nearby swans—will exhibit an even wider range of movements and vocalizations in the social drama of sex.

To look for a potential mate, the swan ventures out again across the lake, but it encounters other territorial swans who hiss loudly or erect their feathers and tuck their heads down in threat displays. It finally finds an available mate, and the two swans exchange an introductory series of rhythmic grunts and snorts. Then they both enter into a ritualized dance featuring slow, graceful movements of their long, flexible necks. Facing each other with bowing necks, they turn their heads in alternating directions, occasionally stretching them up high or rubbing them against their partner or their own breast. Turning in parallel, they synchronously dip their heads down into the water and shake them, elevating them slowly with their necks sometimes entwined or their neck plumage puffed out. If both swans are convinced by these multisensory displays, they will copulate and stay mates for life. These displays are generated largely through the swan's elegant and exaggerated neck that makes up nearly half its body length.

• • •

Biologists usually seek explanations for exaggerated body parts, including necks, in two mechanisms of evolution: natural selection (differential survival) and sexual selection (differential mating success). The logic of these two evolutionary forces is relatively straightforward. In natural selection an animal with an unusual neck that enables it to better obtain food and oxygen while avoiding predators and disease will more likely survive and pass on those neck traits to its offspring. In sexual selection an animal with a neck that allows it to better acquire a high-quality mate will leave more abundant and vigorous offspring and thereby more copies of its own unusual neck form. If either of these forces act consistently over many generations, necks can become exaggerated in any number of dimensions: necks that are long or short, flexible or stiff, loud or quiet, showy or drab.

Although the theory of these two selective forces is relatively simple, the way they act and interact in real animals is often quite complicated. Multiple forces act within both natural and sexual selection, and natural and sexual selection influence each other. For instance, an animal with a neck that uses songs or ornaments to effectively lure mates may also catch the attention of predators, pitting sexual selection against natural selection. An animal with a certain neck ornament that attracts one kind of mate may repel another sort, and so there is no singularly superior neck design. An animal that attracts mates with its vocal repertoire may simultaneously produce a conspicuous visual display at the throat, and it can be difficult to know which feature is subject to selection. This complex interplay of selective forces and how they generate the diversity of necks is the focus of this chapter.

Necks are a site of natural selection because they have many functions crucial for survival: foraging, ingestion, respiration, and predator detection. But necks also participate in reproduction, usually through their role in sexual communication, so they are subject to sexual selection as well. The most familiar of these sexual signals are the elaborate mating calls sung by birds and frogs in their glorious choruses at dusk and dawn. However, sound is not the only means necks can be used for getting the attention of potential mates. In some cases necks are also visually loud. In turkeys and other fowl, males have colorful wattles that conspicuously dangle from beneath their bill. Many lizards have color patches on their throats in the breeding season. Usually these sexual signals are displayed by males to attract females, and females might select among potential mates based in part on these neck displays, though in a few species males choose their mates based partially on female throat coloration. In all these species individuals that are most noticed and chosen as mates based on their neck ornamentation or mating calls pass on their sexy necks to abundant offspring.

Necks seem obviously well suited for acoustic courtship because they contain an air-transmitting pipe, the trachea. But why visual courtship through ornamentation occurs so commonly at the neck is less obvious. One reason might be that the neck is simply near the head. Animals are probably especially interested in the heads of other animals because the head and its gaze often reveal attention and, importantly, returned attention. Moreover, animals often elevate their head to better survey their environment, making them especially noticeable to others. The high degree of mobility in necks further enhances the conspicuousness of heads, since moving stimuli are always more detectable than stationary stimuli. So sexual ornaments may have evolved commonly at the neck because there they are especially visible for communication. Another reason might be that the neck can afford a lot of excess communicative baggage. Most ornaments are composed of extra skin, feathers, or hair. In other body regions (e.g., appendages, tail, or belly) this extra tissue might impede or be damaged by locomotion. But the neck is usually tucked behind the head, where it is mostly out of the way and protected from scrapes or bumps.

Sexual signals are conveniently divided into acoustic versus visual signals, but courtship at the neck is sometimes multimodal, stimulating both audition and vision together. For instance, grouse puff up huge sacs on their throats that simultaneously emit deep rumbles and display colored patterns. In many species olfactory and tactile cues further contribute to courtship. These various sensory channels can work separately to lure mates at different ranges and at various phases of courtship, or work in combination to heighten the overall appeal of a mate.

In addition to these alluring, often beautiful courtship behaviors, necks participate in the aggression surrounding reproduction in some species. Competition for mates, usually among males, sometimes entails combat involving the neck. Male bighorn sheep clobber each other with their horns, which inflicts enormous forces on their cervical spine. Male elk use their necks to carry around huge antlers that function as displays and weapons in territorial disputes. Male wolves growl from the throat during combat. In some hummingbird species dueling males jab their opponents at the throat with their sharp bills. The winners of these contests commonly have greater access to mates and thereby pass on their macho necks to a disproportionately large number of offspring. So sexual selection can act at the neck through such competition for mates as well as through mate attraction and choice.

Biologists have spent decades teasing apart which of these forces—natural selection versus sexual selection, mate competition versus mate choice, and selection on acoustic versus visual sexual signals—or which combination have shaped the unusual necks in the animal world. Below, I describe a collection of animals who use their necks in sexual interactions—lions, giraffes, lizards, and frogs—as examples of how sexual selection can contribute to the phenomenal diversification of animal necks and as stories of how biologists test and argue about their notions of what drives this diversification. Then I turn to our own species, examining sex and gender differences in human voice, anatomy, and display at the neck. Anthropologists and psychologists along with biologists have incorporated evolutionary ideas into their explanations for human biological sex differences. They have also examined how culture influences the ways women and men often ornament their necks differently. I follow definitions of the National Institutes of Health, using "sex" as a biological construct involving anatomy, physiology, and genetics that applies to both human and nonhuman animals and "gender" as a sociocultural construct involving identities, roles, and norms that applies to humans alone.[1]

COMPLEX SELECTION AND SEXUAL COMMUNICATION IN ANIMALS

Only mammals cover themselves in hair. In striking contrast to the colorful skin, scales, and feathers used by other vertebrates in sexual signaling, one apparent limitation with hair is that it is always drab in color: shades of brown, yellow, black, and white. Consequently, mammals do not ornament their necks with bright colors for sexual attraction. Instead, they embellish their necks by making them large and puffy. Certain species of primates (macaques, baboons, mandrills), carnivores (wolves, tigers, lions), ungulates (bison, certain camels, antelopes) grow hair in great volume all around their necks. In most, but not all cases, such manes are most prominent in males, suggesting their function in sexual interactions.

It is easy to distinguish the sex of lions by their necks. Males have manes, and females do not. Over evolutionary time, selection has acted differently according to sex. This much is clear. However, the precise way sexual selection has driven the evolution of the mane is not so simple. It turns out that female choice as well as male competition probably each played a role, and both selective forces were influenced by the intense heat of the savannah.

Darwin posited that lion manes provide protection in fights among males.[2] Male lions compete intensely, sometimes violently, for females, and the extra fluff of a mane might guard their most vulnerable anatomy from puncture by an opponent's tooth or claw. In this explanation the mane functions in survival. But the wide variation in mane shade suggests that it is more about sexual display. Among males within a social group the mane varies from light blond to dark black. Since light and dark manes would be equally effective shields in male combat, Darwin's protection hypothesis does not explain this color variation. Mane length also varies within a population, often appearing longer in higher-ranking males. So a female might choose her mates and/or a male might assess his competitors for mates based on their manes. Thus, within the sexual arena, either males or females could be driving selection.

When Peyton West was searching for a dissertation topic in the late 1990s, she wanted to address a big question. She fell into conversation with lion researcher Craig Packer, who told her that there were only two big questions remaining in lion biology, one of which was, Why the mane? West set out to answer whether the mane was for protection, as Darwin proposed, or for sexual display, as field studies had suggested. With loads of field data accumulated by Packer over decades, a bevy of undergraduate assistants, and a collaboration with a Dutch toy company, West ultimately concluded the mane was a big show put on for sex, and moreover it was a show that males and females watched differently.[3]

Combat among male lions is sometimes vicious, but it is exceedingly rare. So it was difficult to directly test the protection hypothesis. However, using past photographs and descriptions of lions in Tanzania, West and her colleagues found that wounds inflicted by other lions were neither more common nor more lethal at the neck than any other body region. So they dismissed the mane-as-shield hypothesis. In this case Darwin was probably wrong.

Turning toward the idea that manes are for sexual display, West and Packer asked whether manes are a frivolous show or an honest show. That is, are manes arbitrary displays, or do they signal something real about the quality of a male and his likelihood of survival and reproduction? To address this question, the researchers evaluated the injury rates and physiological status of males that varied in mane length and color.[4] Based on decades of photos from the field, Peyton found that males with longer manes tended to have fewer battle scars elsewhere on the body, and following combat, injured males tended to lose mane hair and replace it with shorter hair. Thus, whereas a long mane apparently does not protect against neck wounds, it accurately signals a male's relative resistance to overall body injury and, by extension, his prowess in fighting. In addition, mane darkness correlated positively with age, testosterone levels, and relative body condition. Thus a mane is an honest signal of both hormonal machismo and ability to acquire food, encoding reliable information about the vigor of a male.

At this stage of the research, however, it was not clear if other lions, male or female, even notice the differences among manes. Here is where the Dutch toy company comes in. To experimentally test lions' reactions to specific features of the mane, West and Packer needed a way to separately manipulate mane length and darkness. They enlisted a toy company that normally makes animal dolls to build realistic life-size lion manikins with interchangeable mane wigs, one for every combination of mane length and color: short and blond, short and dark, long and blond, long and dark. When West placed these figures in the field, the lions were fooled, at least initially. Lions approached the manikins as though they were real, but males and females differed in their reactions. Males avoided dark-maned manikins and tended to spend more time near the short-maned manikins. Since the mane is indeed an honest signal of a male's dominance or fighting ability, perhaps in a naive encounter a male might prefer to hang around a less macho male. By contrast, females affiliated most with males sporting darker manes but had no preference for males based on mane length. These experimental results, combined with observation that females typically choose to mate with dark-maned males when they have a choice, indicate that dark manes are sexy as well as honest.[5]

If dark manes are so effective for attracting lionesses, why don't all males have dark manes? Such sexy ornaments might have a physiological cost and threaten survival. A dark mane can make a male hot—in the thermal as well as sexy sense. Many lion populations live in hot climates where males are particularly susceptible to overheating because, with their larger body size, they have relatively less surface area to dissipate heat. Their necks, which in many mammals are sites of net heat loss, are covered with a thick layer of furry insulation. Within this context of potential thermal stress, dark manes absorb more heat and worsen the threat. Indeed, West and Packer found using infrared-sensitive cameras that within a population males with darker manes had higher surface body temperatures than those with light manes.[6] This susceptibility to heat stress might explain why, on the hottest days, dark-maned lions spend less time hunting than light-maned males. In addition, they tend to eat smaller meals, since gorging on meals is known to elevate body temperature. Thus, during intense heat, dark manes might compromise a male's food intake.

Altogether it appears that sexual selection, acting through both female choice and male competition, drives males toward darker manes. However, because of the thermoregulatory costs of dark manes, natural selection opposes sexual selection and maintains part of the population with light-shaded manes. Necks that are sexy in one context can be dangerously warm in another. During periods of elevated temperature, such as those that are likely more frequent in this century of climate change, dark-maned males might suffer greater mortality from overheating or starvation because of their showy necks.[7] By heating up their habitat, humans may be meddling in the interplay between natural selection and sexual selection in lions.

Across the savannah from the lion habitat lives another beast with the most famous of all necks. Giraffes and their long necks (2.5 meters, or 8 feet) beg for an explanation. Their exaggerated shape has long captured the human imagination, and just-so stories from diverse times and places offer a variety of explanations for what caused the original protogiraffe to lengthen its neck. For biologists the classic evolutionary explanation for the giraffe's neck, retold in practically every biology textbook, relies on natural selection. The giraffe's neck elongated over generations because it provided a survival advantage in foraging for leaves high in the trees.[8] Indeed, the giraffe neck is used perhaps more than any other biological structure to illustrate Darwin's idea of natural selection. However, research over the past thirty years has indicated that sexual selection may have also contributed to the evolution of elongated necks in giraffes. Longer necks may have provided males an advantage in competition for mates.

Darwin himself offered an explanation that relied on the survival advantage of a long neck.[9] It goes like this. Among the ancestral short-necked giraffes, individuals within the population varied in neck length. When droughts came and the competition for leafy food sources intensified, those with longer necks could forage at higher levels, and they were more likely to survive than shorter-necked individuals. Because neck length is passed on to offspring, this differential survival resulted in subsequent generations possessing longer necks. Although Darwin emphasized the foraging advantage, he also recognized that a long neck would confer additional survival advantages. For example, the giraffe's long neck would serve as a "watchtower" enabling it to better detect predators or as a weapon to butt away predators. Since Darwin, other scientists have proposed an additional thermoregulatory benefit for a long neck. The giraffe's long neck held high in the breeze might serve as a radiator that could cool it down in hot African environments. Given the multipurpose use of the neck, its elongation would have multiple survival advantages. Darwin and others since him clearly recognized that natural selection could occur on many fronts.

In the mid-1990s Robert Simmons and Lue Scheepers proposed a different, controversial explanation, one based on sex rather than survival.[10] From their direct observations of giraffes in Namibia, they simply did not see giraffes grazing high in the trees. Even during the dry season, when nutritious leaves were scarcest and competition was

Figure 14. Male giraffes sparring through "neck fighting." Photograph by Bjørn Christian Tørrissen, 2015.

greatest, giraffes spent most of the time feeding at shoulder height at a level reachable by other competing species. Apparently, the textbook depiction of high browsing just does not happen frequently in nature, suggesting that while it could have been important in the evolutionary past, it is not in present times. However, Simmons and Scheepers commonly witnessed another behavior that could have an underappreciated evolutionary impact: males use their necks in fighting with each other for dominance and access to females. In the breeding season, dueling males line up in the same or opposite direction and whip their heads against each other with high force. Injuries are common and, in one observed case, the fight was lethal. Male giraffes with the largest necks tended to win these contests, and females were most sexually receptive to males with the largest necks, suggesting that the elongated neck was driven by sexual selection. Males compete for females through combat, and those that win in these "necking" competitions have more access to mates and sire more offspring, who, on average, have longer necks.

A primary objection to this "necks-for-sex" hypothesis was that it did not explain why males and females have proportionately the same neck length, even though only males engage in neck combat. Simmons argued that the long necks of females might be a "neutral by-product" of selection on long neck length in males.[11] In other animals it is well known that although some traits can diverge between sexes evolutionarily, other traits are sexually tied together. Change in one sex requires change in the other sex. That is, the underlying genetic and developmental mechanisms constrain males and females to evolve in the same direction. A familiar example of this phenomenon is that nipples are present in male mammals even though they function only in females. Perhaps the selective pressure for male victory in neck battles was so strong that female neck morphology was "carried along" over evolutionary time. However, without any evidence for such genetically-based sexual correlation, the sexual selection hypothesis for long necks in giraffes initially did not gain widespread favor.

The idea that the long neck of a giraffe arose as a tool in male sexual competition resurfaced in the 2020s based on the discovery of fossilized neck and head anatomy of an early giraffe relative, *Discokeryx xiezhi*.[12] This ancestral member of the giraffe lineage was small—about the size of a sheep—with an elongated neck. Its most distinguishing features, however, were its thick, bony "head cap" and its stout massive neck vertebrae. Both structures indicated that *Discokeryx* engaged in extreme head-butting, with an estimated force that greatly exceeded that of well-known living head-butters, such as muskoxen. Although this head-on form of collision differs notably from the sideward jabs of contemporary giraffes, it demonstrates that from the very earliest stages of giraffe evolution, the head and neck were used as a weapon, probably in mate competition. In subsequent evolution within the giraffe lineage, the further lengthening of the neck and modifications in fighting mode likely proceeded through an interplay of sexual selection for male competition and natural selection for high browsing. While researchers differ in their opinion on whether sexual selection or natural selection was the original or primary force driving neck elongation, most concede that it was probably both.

Neck ornaments are commonly studied along a single dimension: What feature of a neck ornament best predicts the sexual success of an animal, and how is it constrained by natural selection? But in some cases neck ornaments reveal that there is more than one way to succeed in the mating game. Such is the case in the multicolored throats of side-blotched lizards. These lizards are a small-bodied (about 45 millimeters, or 1.7 inches, in body length) abundant species distributed widely through the western United States. Like many lizards, they have colorful patches on their throats during the mating season. But rather than differing in gradations of one color, these throat patches come in three distinct color classes: orange, blue, or yellow. Individuals have only one kind of throat patch color, and each color comes with a different mating strategy.

The evolutionary forces acting on this species and the behavioral interactions of its three-color morphs were the subject of decades of research by evolutionary biologist Barry Sinervo and his collaborators. Orange-throated males are the largest and most aggressive type, defending large territories with many female residents. Blue-throated males have smaller territories occupied by only a single female. Yellow-throated males are "sneakers"—that is, they slip into the territories of other males and sneak copulations on the sly. They can do this, in part, because their yellow throats mimic those of sexually receptive females, so they are not easily detected as competitive intruders. Each morph has its own advantages and vulnerabilities in competition with other morphs. Orange males have access to more mates, but they are stretched thin guarding a large territory, so they more commonly lose mating opportunities to yellow sneaker males. Blue males defend a much more manageable territory, and they even cooperate with other blues to ward off intruders. However, blues only have a single mate. Yellow males do not need to worry about guarding a territory at all, but to mate, they must dodge the aggressive oranges and blues.

Like in a rock-paper-scissors game, each strategy, color-coded at the throat, can be a winner or a loser, and because of this balance, all three morph types persist in the population.[13] Across years, one morph type might increase in abundance, but at higher densities this morph's strategy becomes less advantageous—females tend to prefer the color type that is least common—and eventually another color type predominates.[14] Winners eventually become losers, so the abundance of each morph cycles up and down from year to year. Yet, over time, each color morph sires equivalent numbers of offspring, as evidenced by DNA paternity studies. Diversity of throat color and all its accompanying sexual drama fluctuates with counterbalancing evolutionary forces.

We humans are well aware that in our species courtship is a multisensory experience. Our attraction to potential mates depends on their appearance, their voice, and perhaps even their scent and the way they touch. The courtship signals of some other animals, however, seem so glaring in one sensory modality—for instance, gaudy bright colors or boisterous songs—that we sometimes overlook other sensory modalities involved in the courtship dance. One such case is Túngara frogs, which puff out their throats to make sexual signals that are conspicuous both acoustically and visually. A female frog selects her mate based largely on the precise acoustic details of an individual male's call. Yet these details are embedded with the cacophony of many other suitors. Most frogs generate these songs by inflating elastic vocal sacs in their throats and then pushing this air through their vocal folds. The sounds produced by this vocal apparatus are so conspicuous to humans that until recently scientists have focused nearly exclusively on the auditory channel for understanding mate selection in frogs.

Calling occurs mostly at night when visual signals are hard to detect, at least to humans, and even species that call during the day are usually so camouflaged that they are difficult for us to see. Decades of research have demonstrated that acoustic stimuli alone are sufficient to elicit a mate selection response in female frogs. In an experimental design that has been used in dozens of studies, a female frog is presented with differing frog call recordings through two loudspeakers, and she indicates her preference by moving toward one of the speakers. Such experiments have generated a wealth of information on the predilections of females because the acoustic stimuli can be easily manipulated to allow researchers to identify what specific features of a call are most effective in attracting a mate.

Scientists have recently turned their attention to the role of another conspicuous but usually disregarded feature of frog vocalization: the visual information present in pulsating throat sacs.[15] In Túngara frogs, for instance, the volume of the vocal sacs changes by as much as forty-fold as they inflate and deflate them over a period of less than a second, causing the overall volume of the frog to nearly double.[16] Studying the consequences of such impressive pulsating visual signals is challenging because these visual displays are difficult to mimic and manipulate experimentally.

This changed in 2008 when biologists Ryan Taylor and Michael Ryan teamed up with engineers to develop a "robofrog" that looked just like a real Túngara frog and whose throat sacs could be remotely controlled by an experimenter while the model emitted synthesized sound signals.[17] They used this experimental apparatus to assess the interaction of acoustic and visual stimuli in female mate selection. When they gave females a choice between a male call alone and a robofrog only exhibiting vocal sac movements, females consistently showed a preference for the call alone, indicating that the audio channel is unquestionably the most important stimulus to them. However, researchers also found that females preferred a robofrog whose throat pulsed in time with the call to the sound of a frog call alone. The visual signal boosts the effect of the audio signal. Moreover, when the throat pulsations

Figure 15. Male Túngara frog displaying calling behavior and inflation of vocal sacs. Photograph by Michael Ryan.

were out of sync with the call, females avoided the robofrog. Just as humans cringe at poorly dubbed movies, frogs are averse to mismatched visual and audio communication signals. In their natural habitat in Central America, Túngara frogs must make mating decisions surrounded by the clutter of dozens or hundreds of calling frogs. Taylor and Ryan speculated that a male's pulsating throat sacs help a female associate a particular male with his especially attractive call. The call alone brings the female in the vicinity of a male. Then, at closer range, the visual signal helps her identify which of several nearby males is making her preferred call.[18]

Remarkably, a third sensory modality—vibration detection—also influences perception of an acoustic call.[19] When a male emits his call, his pulsating vocal sacs create ripples in the surrounding water, and a nearby female might be able to feel these good vibrations. To assess the possible role of such seismic communication, biologist Logan James, along with Taylor, Ryan, and others, presented female frogs with a complex call typical of a courting male and gave them a choice between calls that also contained simultaneous pulsating visual signals, rippling vibrational signals, or both. Females preferred the "all of the above" choice; the combination of sound, sight, and good vibrations was the most appealing. The researchers further showed that the right kind of appearance and shake can compensate for an unattractive voice. Females normally prefer a two-part complex call over a one-part simple call. However, when the researchers coupled the simple call with additional visual and vibrational stimuli, it was just as attractive to females as the complex call alone. Pulsing throats and good vibrations make a difference. To complicate matters even further, some species of frogs produce chemical signals in glands in their throat that can be propagated through the water with waves from throat sacs. It turns out that frog courtship, like human courtship, is a many splendored thing, and much of it projects from the throat.

SEXUAL COMMUNICATION AND COURTSHIP IN HUMANS

Like many other animals, humans have necks that differ anatomically by sex, giving us different voices and neck shapes that play a role in courtship and sexual attraction. But unlike other animals, we commonly amplify sex differences by adorning our necks with alluring visual ornaments and dabbing them with seductive scents. Once coupled and in our most intimate sexual interactions, we commonly touch each other's necks, with hands and lips. Our necks send or receive signals in many sensory modalities—visual, auditory, olfactory, and tactile—all in the complex dance of courtship and sex. In our evolutionary past, both natural selection and sexual selection acting on our larynx and neck musculature have likely sculpted these sexual differences in voice and neck shape. While humans have also adopted a vast array of gendered behaviors and ornaments at the neck that have little to do with sex, here I focus on how the neck and its ornamentation play roles in sexual communication and courtship in humans.

Most adult men and women speak with noticeably different voices. From the first utterances of social interaction, our ears distinguish others according to sex. Prepubescent girls and boys have similar voices, and during puberty the voices of both girls and boys lower in pitch and, to some degree, change in quality. Usually, under the influence of testicular (boys) or adrenal (girls) androgens at puberty, the laryngeal cartilages grow larger, the vocal folds grow longer, and both structures vibrate at a lower frequency in both sexes.[20] However, compared to most pubertal females, most pubertal males secrete much more androgen, and their vocal folds become about 60 percent longer and thicker. The larynges of males also typically undergo a "secondary descent" that positions their larynx deeper in the throat and makes their vocal tracts about an inch (2.8 centimeters) longer than those of females. As a result of hormone-driven changes in the throat, the pubertal drop in vocal pitch for boys is much greater, typically falling about an octave in boys, compared to three to four semitones in girls. The difference is so drastic that on the other side of puberty, there is barely any overlap in the fundamental pitch of their voices.[21] The change in the size and position of the male larynx is also apparent visually as a bulge in the throat, the Adam's apple. This male neck protrusion is so visually prominent and characteristically male that all cartoonists need to do to convey the maleness of a character is to sketch a V beneath the chin.

Although vocal pitch and laryngeal size are near-perfect indicators of common biological sex categories, in rare cases the typical criteria for defining biological sex do not align, and this can yield exceptions to sex-typical expressions of voice. For example, sometimes individuals that are chromosomally and gonadally typical of one sex might have a hormone profile more similar to that typically found in the opposite sex. In these cases the pitch of the voice usually aligns with the hormonal profile rather than the chromosomal or gonadal sex.[22]

The human vocal tract was likely molded by natural selection for generating the wide variety of sounds for verbal speech in early evolution. But sex differences in the voice were likely the product of sexual selection for attracting and/or competing for mates. Among apes, humans have likely been especially influenced by sexual selection, since sex differences in human voices are much larger than those of any of our ape cousins.[23] In addition to shaping vocal production, selection has probably acted on our perception of vocal signals, influencing what mates we find sexually attractive and what sexual competitors we find intimidating. As in other animals, it can be difficult to discern whether the evolution of sexual communication in humans is driven more by the sensory predilections of the receiver or vocal capacities of the sender. Like so much in the study of human sexuality, it is the subject of heated debate.

In one view, consolidated by anthropologist David Puts and his collaborators, the low pitch of the male voice is the product of sexual selection as an honest cue of his ability to compete for mates or sire many healthy, fertile offspring.[24] Many studies have demonstrated, albeit inconsistently and sometimes weakly, that men with lower-pitched voices are larger in body size, have greater upper body strength, and are perceived by voice alone to have high dominance by other men—all of which would indicate a high prowess in competition for mates.[25] In addition, when women are presented with audio recordings of men, they rate low-pitched voices more attractive than high-pitched voices, particularly at the midpoint of their menstrual cycle when most likely ovulating.[26] These observations suggest selection favors a deep male voice because of its advantages in attracting fertile mates. Women might find a deep voice in a potential mate more subconsciously alluring because it would predict their own sons' sexual competitive ability or attractiveness. Puts has also proposed that women might be drawn to deep voices because men with such voices tend to have hormone profiles that correlate with strong immune competence.[27] Regardless of the underlying mechanism (male competition or female choice), there is wide-ranging evidence suggesting that voice is indeed a marker of male "success." On the whole, men with lower voices tend to have higher incomes, win elections, have more sexual partners, and sire more offspring.[28] Puts thus argues that vocal pitch was selected for in males because it is an honest indicator of fitness.

Voice researcher David Feinberg and his collaborators disagree.[29] They argue on statistical grounds that the evidence linking vocal pitch to body size, dominance, and immunocompetence is weak. Instead, they propose that the deep male voice was selected for because it "exploited" a preexisting sensory bias for low-frequency sounds in general. Based simply on physics, large objects of any sort can vibrate at lower frequencies than small objects, and our auditory systems pay particular attention to such low-pitched sounds because they can indicate imminent danger from any large source. The low-pitch sound of a person, but also a predator, a falling rock, or a thunderous storm all grab our attention because they are perceived as large and thereby threatening. This particular sensitivity to low pitches is found in most vertebrates and far predates human evolution. Sexual selection in human males might have simply capitalized on this ancient sensory predilection. Deep voices attract because they are most noticeable. Puts and his colleagues countered that such sensory bias might explain the origin of the deep male voice, but evolution would have eventually selected against such "dishonesty" in signaling if it were not in fact useful in competing for or acquiring mates.[30] In response, Feinberg and his collaborators noted that human culture is replete with inaccurate myths linking physical characteristics with behaviors and personalities.[31] The debate continues.

Although the voices of women typically change less than those of men during puberty, there is abundant evidence that sexual selection acting through mate choice has influenced female voices as well. In many women the voice fluctuates according to their reproductive potential, with higher pitches during the ovulatory phase of the menstrual cycle and in younger, more fertile phases of the lifespan. Such changes in pitch likely occur through the action of sex steroids (estrogen and progesterone) on the vocal folds. Women who take oral contraceptives and thereby experience smaller steroid fluctuations show little vocal change.[32] Thus voice is generally an honest signal of fertility. Men appear to take notice. When presented with audio recordings, men on average prefer higher-pitched female voices. Moreover, they rate the voices of women during the ovulatory phase of the menstrual cycle as more attractive than when these same women are in less fertile phases, and more attractive than the voices of other women taking hormonal contraceptives.[33]

These studies on the hormonal regulation and evolution of voice through sexual selection rely on the assumption that our voices are shaped by unconscious physiological changes during our lifetime and selective forces deep in our evolutionary history. But other research shows that humans make situational behavioral adjustments in voice during opposite-sex communication. Voice is something we do, not just have. For example, in one study female college students were first asked to rate the attractiveness of men based on a series of photos and then prompted to leave phone messages asking these men out on dates. In these messages women tended to raise the pitch of their voice, but only in messages to men whose photos they found attractive.[34] There is also evidence for the opposite vocal shift in men. When soliciting dates in a simulated courtship setting, male college students lowered their voice more toward women they found attractive.[35] In these cases the change in pitch was slight, but nonetheless, in such sexually charged interactions, men and women exaggerated their voices in the direction of the sex-typical pitch.

But not all voices follow these sex-typical patterns. Women in positions of power—in both the political and business worlds, for example—sometimes lower their vocal pitch, probably as a way to sound more authoritative or confident.[36] This is often a reasonable choice. Listeners of both sexes generally find a deeper voice in leaders of either sex more confident and authoritative.[37] This deepening of the voice in powerful women was tracked over the political ascendency of British prime minister Margaret Thatcher. In college and her early political career, Thatcher used a high-pitched, "light, airy" voice.[38] Soon after she became Leader of the Opposition, one of her advisers suggested that her voice might compromise her political aspirations. The electorate needed a leader with a stronger tone. With the lauded British actor Laurence Olivier acting as an intermediary, Thatcher received training from a voice coach at Britain's National Theatre. She eventually lowered the pitch of her voice by 46 Hz, placing her voice near the midpoint between the typical male and female voices. It was a voice more suited to her image as the Iron Lady.

More recently, such vocal modulation according to power was evident in the rise and fall of Elizabeth Holmes, the technology entrepreneur who in her twenties led the growth of a $9 billion medical testing company, Theranos. During her company's rise, Holmes used an especially deep voice in public appearances, including in a widely circulated 2014 TED Talk, delivered near the height of her career, in which she pitched her ambitious vision for how her company could transform the whole health-care system.[39] By 2018, in her mid-thirties, Holmes was indicted for her role in defrauding investors and eventually sentenced to federal prison. By the time of her conviction, she had abandoned her deep voice. From an interview Holmes gave for the *New York Times*, reporter Amy Chozick described Holmes's voice as "slightly low, but totally unremarkable" with "no hint of the throaty contralto she used while running her defunct blood-testing start-up Theranos."[40] In the interview Holmes spoke openly about how she created her Theranos persona so that she would be "taken seriously and not taken as a little girl or a girl who didn't have good technical ideas."

With our voices we woo, flirt, intimidate, reveal, and conceal, all in the ornate dance of sexual attraction and competition. In this complex drama the words conceived in our brains (and hearts?) matter, but the mere sounds of our throats are enough to convey our sexual identities and sometimes our intentions. Sex has influenced our voices both in deep evolutionary time and in the quick vocal gestures we project to each other daily. While human culture is filled with verbal pronouncements of love and desire, so much of courtship, at least in its initial stages, is indirect and subtle. Flirtation is the art of suggestion, orchestrated ambiguity. Much of this occurs in the silent sign language of courtship. Humans, particularly women, use the neck in the nuanced visual signaling of courtship: baring the neck, giving a coquettish tilt of the head, or decorating the neck with alluring ornaments.

Necks of women and men are visually distinguishable, even when they are unadorned, because they differ in proportion. In most cultures long necks are considered more feminine, so one might expect the length of the neck to differ between the sexes.[41] It does, but just barely. The average neck length of women is in fact 0.5 centimeter (0.2 inch) shorter than that of men, but after correcting for height, it is 1.3 percent longer.[42] Still, it is doubtful this length difference is even perceptible. The biggest sexual difference in neck shape is in girth: the necks of men, on average, are 5 centimeters (2 inches) greater in circumference and 50 percent larger by volume than in women. Whereas less than one-quarter of neck volume is muscle in women, almost a third of the neck volume is muscle in men. So when we cue in on neck form as an identifier of sex, we are perceiving mostly differences in shape rather than length, but the relative thinness of the female neck creates the illusion that it is longer.

The static image of a long, narrow neck can alone convey femininity, but neck movements also frequently have sexual connotations. Women commonly lift, lower, or tilt their heads in courtship interactions. Indeed, researchers making "field observations" of young women in a singles bar reported that head/neck movements were second only to facial movements (e.g., glances and smiles) in the frequency of "nonverbal solicitation behaviors." In particular, women commonly exhibited "head tosses" (a brief up-and-down movement, less than 5 seconds) or more protracted "neck presentations" (a roughly 45-degree head cant combined with downward movement to broadly reveal the side of the neck).[43] The dance of courtship often begins above the shoulders.

Although neck shape and movement commonly differ by gender, the extent to which we expose our necks and the area just below is perhaps even more divergent by gender. In the vast majority of formal portraits in Western painting over the past five hundred years, women's necks are exposed while men's necks are clothed. In contemporary formal occasions where fashion is on full display (think about the Oscars), women typically bare all from chin to collarbone or beyond while men usually conceal everything below the Adam's apple with collars and ties. Even in the most seemingly utilitarian shirt, the athletic T-shirt, women's collars usually dip lower than men's (even though men are far more likely to exercise without any shirt at all). It's not clear whether women are more apt to display their lower neck because it draws attention toward their cleavage, acting as a sexual attractant, or because it exaggerates the length of the neck, a metric of feminine beauty unto itself in Western cultures. Perhaps both explanations apply. Conversely, most men seem particularly averse to neck exposure. Maybe they are more eager to hide this vulnerable area or are simply displaying in the opposite direction of feminine display: masculinity as nonfemininity. Regardless, sexual and gender identity are there at the neck for everyone to see in both the anatomy and the collars that surround it.

Compared to verbal courtship, visual courtship is usually less declarative and less able to tell a story. However, in the neck ornaments of Zulus in South Africa the courtship signals are visual and nonverbal, but their use is so elaborate that they have many of the same elements of language—both syntax and semantics.[44] Zulus have a complex culture of beadwork, which decorates attire worn all over the body.[45] These beads vary in color, pattern, and sequence, and each of these elements can convey a meaning or can be strung together into a story. In the case of neck-born "love-letters," or *incwadi*, this bead language communicates information that is especially relevant to courtship: whether someone is single or married, available or committed, wealthy or poor, or living nearby or faraway. Beads may also express a range of romantic emotions and stories, from true love to jealousy to heartache.

Making these beaded courtship tokens is mostly the domain of unmarried women and adolescent girls, who often work together and teach each other the craft and language of beads. The early stages of courtship commonly begins when a girl offers a prospective boyfriend a simple string of white beads, an *ucu*.[46] By wearing the *ucu* around his neck, the boy accepts her overture. The girl might then send a sequence of small pendant panels, *incwadi*, that are usually about 2–4 centimeters (1–2 inches) square and dangle from a necklace over the throat. These beaded love-letters communicate more detailed messages, encoded in an array of designs and colors. For example, an upward-pointing triangle signifies an unmarried woman, and a downward-pointing triangle means an unmarried man. Two triangles joined to form a diamond is a married woman, while two triangles touching at their points in an hourglass shape is a married man. Each color expresses two emotions, either positive or negative. For instance, red can mean desire or anger, blue can mean fidelity or hostility, green can mean contentment or discord. Color combinations commonly indicate where someone lives. Designs combined with color take on further specified meanings: a green ring can mean the girl is young but accepting of her boyfriend's proposal; a black ring signifies she is ready to marry.[47]

Beads can convey in their combinations and sequences a remarkably specific message. In one account a woman makes an *incwadi* for her fiancé, who had to move to the city for work, laying out a series of color patches in an intentional symmetrical order—black, red, yellow, blue, yellow, red, black—all arranged on a white-beaded background. Loosely translated, this could be read, "We are to be married, but my heart is aching because our love seems to be withering away. When will you return?"[48] In such codes communication can occur implicitly without the vulnerability of explicit verbal expressions. Although these *incwadi* are commonly dense with some sort of meaning, there is hardly a single language of beads. Meanings vary by place and time, and individuals might develop their own "dialect" through which they can send private messages. So, while *incwadi* displayed on the neck announce to all that a person is somehow engaged in the dance of courtship, the intended meaning is often a matter of personally coded interpretation.[49] Even within the highly developed visual culture of beadwork, courtship, desire, and commitment retain a strand of mystery.

• • •

In Western culture choker necklaces are commonly associated with sexuality, but they originally may have served for protection and concealment. In the ancient world—for example, in Mesopotamia and Egypt—necklaces worn snugly around the throat held amulets to ward off illness at this particularly vulnerable body region. Some women in the goiter belt of the Alps wore broad necklaces around their throat to hide their protruding goitrous bulges. In the mid-nineteenth century a wave of choker fashion was set off by Alexandra, Princess of Denmark, who in fact wore her broad jeweled chokers to conceal an unsightly scar.[50]

In the modern period chokers became associated with sexual display and also violence. They can imply a collar as means of sexual dominance and control. "The neck is both a highly vulnerable part of the body and, in women, one of the focal points for beauty and sexual desire ... [T]he neck-enclosing jewelry that became fashionable in the nineteenth century and was known as a 'choker' explicitly alludes to the asphyxiation that its visual form implies," writes art historian Marcia Pointon.[51] In this era prostitutes in Europe were commonly identified by a close-fitting necklace. This pairing of clandestine erotica and neck ornamentation was elevated to great controversy in Édouard Manet's 1865 painting of a prostitute, *Olympia*.[52] In the painting, Olympia reclines in a pose previously struck in depictions of classical goddesses. She gazes confidently back at the viewer while wearing nothing but a red orchid in her hair, a gold bracelet on her wrist, and a black ribbon tied closely around her neck. She was nude, erotic, and shameless, and at the time that was nothing short of scandalous.

In the late twentieth and early twenty-first century, chokers have often risen during rebellious trends in popular culture with loosening of

Figure 16. Queen Alexandra wearing a choker necklace, a fashion she began to popularize while Princess of Denmark, 1887. Photograph by Stanislaw Walery.

sexual mores. For example, chokers were popular in the hippie, punk, grunge, and goth waves of fashion. Recent research indicates that chokers continue to connote female sexual freedom and promiscuity.[53] Women displaying chokers in photographs were perceived by both women and men college students to have greater "sociosexual orientation"—that is, willingness to have sex outside committed relationships—and women who self-reported wearing chokers rated themselves higher on the scale of sociosexual orientation. Chokers rest in the uneasy place where danger, rebellion, and erotica all meet.

• • •

The voice and appearance of the neck are effective in broadcasting sexual signals over long distances, but at close range humans also communicate sexually with scent and touch. Women spritz themselves with alluring perfume on the neck, and men wear cologne as a sexual attractant. Surely the fragrance industry advertises the seductive power of their products. While scents might be worn at other body regions, the neck is particularly suitable because the heat radiating from its abundant and superficial blood flow causes the perfume to volatilize readily. Moreover, its high position on the body enables the perfume to emanate near the recipient's nose.

As the courtship dance progresses, mates begin to touch the neck, perhaps initially with a sensual massage of the back of the neck and, as the desire heats up, erotically and tenderly kissing the neck and throat. Lovers neck. Here, "neck" is a verb, and "necking" is one of the first things couples do in the transition from distant, uncommitted attraction and affection to corporal entanglement and vulnerability. The neck is usually naked, and its thin and well-vascularized skin often blushes in sexual excitement. It is endowed with dense nerve endings and located near the equally sensitive lips of our lover.[54] Kissing the neck can be rapturous as well as delicate. Often young lovers cannot restrain from the temptation to add a suck to their kisses, bursting the abundant underlying capillaries and leaving a slowly fading emblem, a "hickey," of their budding intimacy. All this touching is a tender pleasure and compelling thrill, yet the neck is a region of such vulnerability that we offer it only to those we have let pass through the nuanced, multistage, multisensory sieve of courtship.

• • •

While necks have played a central role in sexual communication in both our evolutionary and cultural past, humans also display gendered behavior at the neck that has little to do with sexual attraction or competition. Women might sometimes perfume or decorate their necks for seduction, but often they do so to communicate their aesthetic taste or to delight their own senses. Men might wear fragrances or necklaces that are for individual self-expression or pleasure, though these displays are typically quite different than those of women. Work by evolutionary biologist Richard Prum has revived the argument, originally made by Charles Darwin, that aesthetic impulses, such as those for beauty and pleasure, are not distinct from sexual selection but rather its natural spin-offs.[55] Ornaments that may have once served in mate choice and attraction evolved further elaborations because they were simply pleasing to the senses.

One of the most well-known cultural modifications of the human body—neck elongation in the Padaung women of Myanmar and Thailand—is an example of neck ornamentation that appears rooted in gendered aesthetics and cultural identity rather than mate selection or attraction. Many girls and women in the Padaung tribe wear brass coils around their neck, typically extending from the collarbone to the chin. Beginning as young as five years old, girls add coils year by year, reaching up to twenty-five to thirty coils that rise as high as 35 centimeters (14 inches). This series of coils gives the appearance of a greatly elongated neck. However, X-rays show that the coils push the collarbone downward and elevate the chin, giving merely the *illusion* of an elongated neck rather than actually stretching the cervical spine.[56] Although this custom is gender-specific, practiced only by women, these coils appear to play little role in courtship.

There is surprisingly little scholarship on the early history of this practice, and its origins are still conveyed mostly in legends.[57] Women within the tribe have differing versions. According to one legend, women wore these coils to shield their most vulnerable anatomy from attack by the nearby tigers, which were known to seize their prey at the neck. Neck coils continued to serve as a symbol of protection. "Our grandmothers would allow us to touch their 'armor' when we were ill.... One should touch them only to draw on their magic—to cure illness, to bless a journey," writes Padaung author Khoo Thwe.[58] Some women see the neck coils as a link to their feminine origins and as emblems of feminine beauty. The Padaung is a matriarchal culture, and the tradition of neck coils connects them to the "memory of [their] Dragon Mother"—a long-necked female who mated with a male human/angel hybrid and conceived the ancestors of the tribe. "I like the rings because they are beautiful and because my mother wears them," explained a twelve-year-old girl.[59] An older woman in the tribe reported, "It is most beautiful when the neck is really long. The longer it is, the more beautiful it is. I will never take off my rings. I'll wear them until I die."[60] Sometimes practices that differ along gender lines are simply a matter of heritage and aesthetics.

While the origins of the neck rings are uncertain, the practice in the past decades has become entwined with economics and cultural voyeurism.[61] To varying degrees, Padaung women have been put on display as curiosities since the early twentieth century, when individual women were "loaned" to officials in Thailand or flown to Europe as spectacles in circuses. Such voyeurism expanded in the late twentieth century as the Padaung were displaced from their homelands following a campaign of ethnic cleansing in Myanmar. Pressed into a marginal existence, scores of Padaung women became part of a tourism industry in which foreigners could pay businesses to be transported into Padaung settlements to view the long-necked women. What may have originated as ornaments of beauty became hijacked as oddities for gendered economic exploitation.

• • •

Sex and reproduction are necessarily whole-body processes. The gonads and sex organs in the lower half accomplish most of the direct acts of reproduction, while at the other end the head collects diverse sensory stimuli to inform the discerning brain when and with whom to mate. The neck transmits nerve impulses and blood-borne hormones to help link these two reproductive processes. Beyond this, the neck appears to play little direct role in reproduction. Yet on second thought, it becomes apparent that the neck plays a vital and indispensable ancillary role. It is a long-distance attractor. The voice broadcasts sounds that draw in mates. Neck ornaments—both natural and created—grab the visual attention of suitors. And in some species, notably humans, the neck surely acts in short-distance bonding; it's an area of such vulnerability that it is only entrusted to intimate partners. So, while the neck internally houses the nerves and vessels for physiological communication between the head and body, it also sends messages out into the world to make the necessary social connections for reproduction. Within and between bodies the neck is a sexual go-between.

CHAPTER EIGHT

Membership & Status

Signaling Identity at the Neck

In the twenty-first century the United States has elected several unusual presidents. In 2008 and 2012 we elected a Black man from Hawaii with—to use his words—"a funny name." In 2016 we elected a billionaire businessman and reality-show star with no political experience. Understandably, some of the electorate were not sure what to think. Who were these candidates? What was their character and allegiance? To answer these questions, some journalists sought clues at the neck.

A 2008 *Newsweek* editorialist carefully read the neckties of the presidential candidates to probe their identity as men and as leaders.[1] Candidate Barack Obama sported a tie with a four-in-hand knot, "an awkward and asymmetrical cinch invented by 19th-century carriage drivers (who held four reins in hand)." He displayed "a knot for the masses." By contrast, Obama's opponent, John McCain, wore a symmetrical, triangular Windsor knot, which "screamed old-guard Washington establishment." Given that there are more than a hundred possible variations on tie knots, why did Obama choose this "pedestrian" version when McCain chose a far more "elegant" and "presidential" knot? It must mean something. The author proposed theories in five categories: biography, rebellion, aspiration, politics, and physics. Decoding neckties is clearly not a simple matter.

In 2017 writers at *Newsweek*, along with those at the *New York Times* and the *Boston Globe*, published equally long articles speculating on the meaning of Donald Trump's unusual neckties.[2] Writing as a guest editorialist for the *New York Times*, Richard Thompson Ford was shocked at the unconventional and seemingly inept necktie habits of Trump. Claiming that the "knot in one's necktie is a biography in silk, communicating details of temperament, character and upbringing," Ford noted that Trump's predictably "bulky Double Windsors are brash and extroverted." But for Ford it was the tie length, rather than the knot shape, that most required explanation. Trump "makes the front end much too long—it hangs far below his waistline—while the narrower end sits, stubby and forlorn, only inches below his collar." Furthermore, as the worst of all transgressions, Trump fitted his tie with cellophane tape—even at his inauguration—to hold the short end in place. Ford speculated that Trump's overly long ties might be a trademark rejection of the "sensibilities of the elite," ultimately concluding that "Mr. Trump's tie symbolizes one of the central questions of his candidacy, and now his presidency. Is his seeming ineptness genuine? Or is it part of a contrived performance, designed to deploy the symbols of power while rejecting the conventions of civility that have traditionally defined and constrained them?" Both the press and Trump have thrived on such questions of ambiguous presidential identity.

For the past several centuries in the Western world, the neck has been one of the few locations where men express their fashion taste in formal attire. The wide variation in neckties is one glaring exception in the otherwise highly constrained world of men's fashion (constrained at least in comparison to women's fashion). Formal suits are usually limited to plain black, gray, or blue; shirts are uniform in color, or perhaps striped or plaid; pants are either pleated or not. Against this drab and uniform background, ties range profusely in style. For example, tie width varies by decade: nothing dates a photo more quickly to the early 1960s or the mid-1970s than neckwear. Compare the narrow ties of Presidents Kennedy and Johnson to the wide ties of Nixon and Carter. Tie color, pattern, and knot shape are even more diverse and are the subject of lengthy commentary in evaluating the fashion tastes of male TV personalities as well as presidential candidates.[3] But beyond conveying individual taste, neckties, like other forms of neck decor, have been important indicators of group identity and social standing. This chapter elaborates the many ways that such nonverbal visual language projects from the frame below our face.

• • •

One peculiarity of the neck is that visually it is usually quite public. We usually bare our neck for all to see, and often we decorate it to draw in visual attention. A common effect of many such decorations—ties, scarves, collars, jewelry, tattoos—is to display our group or self-identity, our membership or rank. If our face is our portrait, our neck is often a caption beneath, conveying social standing or individuality.

Of all anatomical regions, why do we use the neck as a site to communicate identity and social status? Probably because it is both conspicuously located near the face and narrow in shape. We all want to know the group affiliation and status of others, but, importantly, we do not want to appear too eager to know. Necks and their decorations are visible even when the viewer is pretending to attend only to the eyes and face. You can see the portrait and the caption in a single glance. Imagine if we wore our badges of identity on our feet or waist. You could easily catch an identity-gawker. In addition to its placement within polite gaze, the neck is a constriction that makes it a convenient location to secure and remove emblems of identity. The neck is an anatomical signpost. We can easily tie, clasp, wrap, and hang objects there. At the neck such emblems are out of our own way but in the sight line of the public.

Our species is not the only one to signal membership and status at the neck. Many terrestrial vertebrates have elaborations at their throats and necks that indicate their species identity or social rank. Perhaps even more than in humans, such signals in animals often function in courtship and other sexual interactions, as discussed in chapter 7. Here,

I focus on cases in which these signals play a role in social interaction beyond mate selection. For example, some lizards flash colorful throat patches that are more for broadcasting species identity than for attracting a particular mate. Some sparrows have throat plumage that conveys social rank and age, even outside the breeding season. While identity and status signals in humans rise and fall through historical periods and fashion trends, such signals in animals have arisen through eons of selection to be conspicuous and communicative in the context of complex animal societies.

GROUP IDENTITY

While presidential ties have been scrutinized for clues about the individual characters, neckties of everyday men have been used as markers of group identity. Their colors or patterns can signal loyalty to an alma mater, belonging to a family clan, or origin in a particular geographic region. They might announce membership in a club or service in a particular branch of the military. They project, usually with great pride, affiliation with an institution or subculture. Neckties and other neck ornaments are so effective in conveying identity because they are succinct as well as eye-catching. To the intrigue of anthropologists, ties illustrate a notion generalizable all over the world: that communication often relies on arbitrary signals. Symbols of group identity—flags, mascots, logos—seldom announce group membership in words. Rather, they serve as minimal abstract visual codes and little else.

Anthropologist Meyer Fortes argued that neckties have so much communicative possibility because they have no function beyond communication.[4] Fortes wrote about his observations in the 1950s and 1960s of the near-universal practice among men at British universities sporting neckties that marked affiliation with a college, club, or social class: "The necktie, in short, is an ideal badge of group membership and loyalty, be it in an organised association, or in a casual clique, or merely among men with common values.... And it is ideal for these purposes because the message it conveys is not confused by the requirement of utility." Most clothing must cover, protect, insulate, or carry things. Ties are pure fluff, pure communication.

In the West, neckties have served for centuries as "useless" signals of group membership in formal social occasions, but they probably originated as a form of national identity that served a crucial function on the battlefield.[5] In the first military uniforms worn by whole armies, soldiers during the Thirty Years War donned colorful sashes that enabled them to distinguish friend from enemy. For example, in the Battle of Lützen (1632), the Catholic Imperial troops wore red sashes near the neck and the Protestant Swedes wore green. Such brightly colored signaling may have been particularly useful since this battle is legendary for its dense fog. The cravat also has military origins. When the French were fighting the Austro-Hungarian Empire to wrest control of Turkey in 1678, they hired Croatian mercenaries who adorned their necks with colorful silk bandanas. Later, as these Croatian troops marched victoriously into Paris, Louis XIV was so delighted that he began wearing similar neckwear and popularized it in the civilian world. He designated a whole regiment of his army as the Royal-Cravates. Etymologically, cravats derive their name from these mercenary Croats. The bandanas Croats wore around their necks were themselves probably modeled from clothes worn by Roma living in peripheral settlements of Croatia. How ironic it is that neckties, which now are such symbols of establishment power, may have roots in the ornamentation of marginalized Roma.

In contrast to neckties, scarves worn at the neck have a broad range of utility. In both sexes scarves insulate the neck for warmth. In some cultures and certain historical periods, scarves in women's attire concealed the neck and upper chest in modesty and sometimes wrapped upward to cover their head or protect their coif. But like neckties, scarves have been used as symbols of group identity. For more than a century women have worn scarves to signify solidarity with key struggles in women's rights.[6] In the early twentieth century the Women's Social and Political Union in England distributed scarves in their colors of green, purple, and white for members to signal their support for women's suffrage.[7] "Purple ... stands for the royal blood that flows in the veins of every suffragette ... white stands for purity in private and public life ... green is the colour of hope and the emblem of spring," explained Union founder Emmeline Pankhurst.[8]

Decades later, women in Argentina wore white scarves as they gathered to protest the disappearance of their adult children under the dictatorship of Jorge Rafael Videla. In this case white was a symbol not of purity but of motherhood. In 1977, at the inception of the movement that was later called the Madres de la Plaza de Mayo, founder Azucena Villaflor de Vincenti suggested that women identify themselves as part of the protest by wearing white cloth diapers on their heads and tied around their necks, "because every mother keeps something like this, which belonged to your child as a baby." For years afterward, hundreds of women gathered every Thursday in the Plaza de Mayo of Buenos Aires wearing these diaper scarves, many of which were embroidered with the names of missing children. When Argentine women began organizing the National Campaign for the Right to Abortion in 2005, they drew on the emblem of the Madres's scarves but transformed them to green bandanas, as a "symbol of nature, growth and life," and tied them around the neck.[9]

By 2018 thousands of abortion rights activists—"a sea of green"—gathered outside of the Argentine Congress to support a bill decriminalizing abortion. The emblem was so popular that the country started running out of green fabric. Meanwhile, abortion opponents began wearing blue bandanas around their necks in counterprotests. The abortion rights movement and its green neck emblem spread across Latin America, and when abortion protections guaranteed in the *Roe v. Wade* legal case were overturned by the US Supreme Court, crowds of protesters adopted the same symbol. Today the green bandana, especially draped around the neck, is well recognized as an international emblem of women's reproductive rights.

• • •

While ties and scarves commonly display group identity, our language about collars—specifically their color—encodes occupational identity.[10] Appearing in the 1910s, the term "white collar worker" referred to managers and clerks in clean offices shuffling papers with little fear of soiling their clothes. Historically these workers usually wore white shirts and were wealthy enough to hire someone to bleach and press their shirts. The term "blue collar worker" arose a decade later to refer to laborers wearing clothes dyed in dark colors to cover up the grease, dirt, and grime that smeared them in their daily manual labor. In principle, such work clothes could be most any dark color. Why blue? Author Jude Stewart explains that blue dye is the link between the durability and the comfort of laborers' clothes.[11] For durability, workers' clothes must be thick and tough, but this also tends to make them stiff and uncomfortable. Indigo dye helped make stiff cloth pliable. Unlike most dyes, which penetrate the threads, indigo dye sticks on their surface. When indigo-dyed cloth is repeatedly worn and washed, the dye flecks off and takes some of the fibers with it. This causes the fabric to fade but also to soften, a process familiar in our own favorite blue jeans. Beginning in 1911, Levi Strauss's company began dyeing its classic work pants with indigo. While denim was still too bulky to make a comfortable shirt, blue jeans were often paired with blue shirts made from a softer, paler variant. These blue shirts and their collar that wrapped the neck thereafter became the namesake of manual workers.

Beginning in the 1970s, collar color as an identifier of occupational status underwent a great diversification. In her 1978 book *Pink Collar Workers*, Louise Kapp Howe argued that despite much talk of women's equality, the workplace was still greatly segregated by gender, so much so that entire classes of jobs—secretaries, nurses, elementary school teachers, bank tellers, flight attendants—were populated nearly exclusively by female workers, who became known as "pink collar workers."[12] By the twenty-first century, collar designations spanned the full spectrum of the rainbow: gold collars for "brainpower workers" in finance, technology, medicine, and law; green collars for workers in environmental jobs;

orange collars for prison workers; and scarlet collars for sex workers.[13] There are now even "new collars" for an emerging class of workers that require advanced skills but not advanced degrees and "virtual collars" for robots, who might replace certain working classes all together.[14]

At the same time that collar designations have multiplied, the traditional constraints on business attire have started to relax, and some businessmen have readily shed their power tie and collar altogether. The most famous executive associated with no tie and no collar was Apple Inc. founder Steve Jobs. Founding a company whose brand was closely identified with its signature design and style, Jobs styled himself as a fashion disruptor by eschewing the traditional neckwear. He was a turtleneck man, specifically a black turtleneck: no collar, no color. He wore antibusiness attire. In Apple's products Jobs championed the idea of controlling highly complex machines with minimalist icons, and in his personal image he projected a minimalist visage, with short hair, short beard, and rimless glasses. His minimal neckwear seemed to further declare his liberation from unnecessary clutter. His head rose high above the whims of ties that signal caste and collars that signal occupational status. In the subsequent burgeoning technology establishment, tech giants 2.0 (including Sergey Brin, Larry Page, Elizabeth Holmes, and Mark Zuckerberg) continued to broadcast their pretension of antiestablishment values by shunning establishment neckwear.

While the dress code for the neck in the formal business world appears to have loosened, the religious establishment appears to be holding on to at least one widespread tradition. Almost all formal religious attire, worldwide, covers a large portion of the neck. Such religious neckwear conceals the body below the face in modesty while exalting the head in authority. It delineates a clear boundary between head and body, and as mentioned in chapter 1, several Western religious traditions conceive of the neck as separation between the purer operation of the head and the more profane function of the body.

Since the early twentieth century, the unmistakable badge of the Christian clergy has been the clerical collar, a stiff black collar surrounding the neck with a plain white square in front. Christian clerics began wearing distinctive dress in public in the sixth century, and after the thirteenth century it was common for them to top their black robes with white linen collars. For a period this collar became more ornamented and lacier, until in 1624 when Pope Urban VIII prohibited such extravagances. The modern, removable clerical collar was invented in the 1890s, and its widespread use by World War I chaplains solidified it as the near-universal symbol of the Christian clergy.[15] Even in more secular contemporary times, these collars elicit respect. A simple white square can prompt people to put away guns, confess their sins, stop shouting, speak more properly, or just straighten their unkempt hair. The clerical collar still commands a certain moral authority.

Clerical identity can bring respect, but it can also bring danger. In 2007, for example, Father Paul Bennett of South Wales was stabbed six times in the back by twenty-three-year-old Geraint Evans, who was "fixated on religion and knives" and declared himself Satan.[16] In response, a subsequent report from the Church of England urged clergy and even the Archbishop of Canterbury to adopt more "casual" clothing because the clerical collar made them an "easy target."[17] Following a separate 2016 incident, British counterterrorism specialists warned priests about wearing their clerical collars after two priests conducting Mass in their churches were stabbed by ISIS-inspired jihadists.[18] Sadly, religious neckwear can summon violence as well as inspire reverence.

My father was a dedicated churchgoer and took his kids, mostly compliant, every Sunday. He had a high tolerance for sloppy dress, sandals, long shaggy hair, and other irreverent exteriors. But there was one limit. If a razor had not touched the neck of his sons in several days, his big forefinger would swipe at the sides of our necks: "Why don't you see if you can trim that up a bit." For a remarkably understated man, those were as demanding as his words got. He was fine with a bushy beard, but he could not stand a fuzzy border separating it from the neck. Decency has limits and so does the neck. I never considered my father prescient in the ways of pop culture, but perhaps he anticipated the emergence of a forthcoming, twenty-first-century symbol of decadence: the neckbeard.

According to the website "Know Your Meme," the neckbeard as an internet icon entered popular culture in 2003: "Neckbeard is a pejorative term referring to unattractive, overweight and misogynistic Internet users who wear a style of facial hair in which a majority of the growth is present on the chin and neck."[19] Another website portrayed an "overgrown guy in his late twenties that smells of Cheetos and Mountain Dew and spends most of his days in his parents' basement playing video games." Hardly a flattering identity. Why does hair on the neck excite so much revulsion that it provides the namesake of such a repulsive character? Is it a marker of immaturity? Are they too lazy to attend to the most basic grooming? Or perhaps we are disturbed by a fuzzy line at the borderlands between our head and body? Regardless, to this day I can barely let my neck whiskers grow to more than a stubble.

"Rednecks" no doubt get their name, in part, from their work. This social class is commonly associated with rural areas where many people work long hours in the sun, turning their necks red. But "rednecks" are far more of a culture than an occupational class. Farmers in California or ranchers in New Mexico may very well have deep red necks, but they are probably not "rednecks." Historically associated with Appalachia and the Ozarks, "rednecks" carry a defiant pride in their insular, lower-middle-class identity. There are many explanations about where and when this distinct American icon came to be known as a "redneck." Some trace its origins to the lower Mississippi Valley, the swamps of Arkansas, or the sharecropping farms of Georgia.[20] According to one theory, this moniker may have deep roots in old Europe. In this explanation the necks of the original rednecks were tinted red not by rays of the sun but by the blood of rebellious Presbyterians.[21]

In the 1640s, Presbyterians in Scotland defied the hierarchy of the ruling Anglican bishops and declared their desire to separate from the Church of England by signing documents of rebellion in their own blood. To sustain the movement, many so-called Covenanters wore strips of red cloth around their neck to commemorate the bloody declaration of independence. The descendants of these rebels spent several centuries in the lowlands of Ireland, but in the late eighteenth and early nineteenth century, many migrated to the United States. By this stage in the colonization of the United States, much of the prime agricultural land along the Atlantic coast was occupied by earlier immigrants, and these Scots-Irish tended to move farther west into Appalachia, where they commonly settled in isolated "hollers." Yet through centuries of migration they retained their defiant attitude and identification with red color at the neck.[22]

• • •

Most badges of identity can be modified according to the situation or removed altogether. We all take off our neck-borne badges—ties, scarves, clerical collars—when we retreat into our domestic life, where no one needs to be informed of our identity or impressed with our style. But one form of badge is indelible. Tattoos are forever. Most neck tattoos, because they are difficult to conceal, are always and for everyone. It takes one level of commitment to engrave an image on your shoulder or hip; it is permanent but with a restricted audience. It is something altogether different to display your inner dragon or your romantic allegiance in that public space above the collar. Such permanent and conspicuous markers of expression and loyalty are worn for a lifetime, eliciting questions, admiration, or judgments from onlookers.

In some cases, when neck tattoos are used as badges of group identity, the consequences can be far greater than a curious or shaming stare. When Robert Torres was fifteen years old, he got a small tattoo of the letters "SF" on his neck as a display of his loyalty to his San Fernando Valley gang, the San Fers; it eventually cost him his life.[23] Soon after getting his tattoo, Torres had several stints through the prison system, where his tattoos caused him to be shunted automatically to the gang unit. Before his last release, he resolved to "come clean" and return to his ex-wife and five children. On his second day out on the streets, however, he was fatally shot by a member of a rival gang, the Shakin' Cat Midgets. It was payback for a drive-by shooting by the San Fers that occurred weeks before Torres's release. The assailants identified Torres's gang membership by his neck tattoo.

Economists Ray Fisman and Tim Sullivan sought to explain the logic behind such potentially lethal badges.[24] They reasoned that running an organized crime organization requires a lot of quickly accessible information, and among the most important pieces of information are membership and sincerity. Signaling allegiance to the group with mere words or with removable or concealable symbols—clothing, neckwear, or hidden tattoos—is cheap. Submitting to a neck tattoo is the ultimate symbol of commitment because it is public and permanent. Moreover, by sabotaging your employment prospects, a neck tattoo signals you'll never go mainstream. "Such tattoos make life outside gangs so hard that they serve as an effective commitment not to defect; as Robert Torres learned, it's too costly to live on the outside."

• • •

We humans go to great lengths to communicate our group identity to other humans—that is, to other members of our own species. But we rarely feel the need to communicate our identity as humans to other species, and we never confuse humans with nonhumans. However, animals that live among similar, closely-related species must somehow broadcast their species identity so that they do not bother with the futility of mating with the wrong species. For some animals, such species signaling emanates from the neck. The most famous of these is a collection of lizards in the New World tropics.

If you sit for long in many spots in the West Indies, your attention will likely be piqued by a nearby neck: an anole lizard will probably be flashing its brightly colored throat skin, its dewlap, while bobbing its head erratically up and down. Anoles are one of the most species-rich groups of all terrestrial vertebrates, and this diversification is related, at least in part, to communication signals from their flashy necks.[25] Almost

Figure 17. Anole lizard displaying dewlap, a colorful patch of skin flashed under the throat used in species recognition. Photograph by R. Colin Blenis, 2010.

all species in the genus *Anolis* have a modified hyoid apparatus consisting of cartilaginous bars that are hinged at the front and extend rearward, running just underneath the throat skin. When they display, they swing these cartilages downward to expose a patch of colorful skin below their jaw and neck. Such displays are integral to the lifestyle of these lizards.[26] For example, during the breeding season, males devote an enormous part of their days to making dewlap displays—in some species as much as 95 percent of their active time, flashing their dewlaps up to one hundred times per hour. These displays serve multiple functions, including announcing their presence on their territories and warding off intruding males. But here, I focus on their importance for signaling species identity.

Among the approximately four hundred species in the genus *Anolis*, dewlaps vary dramatically in size and color, and head bobbing varies by rhythm, duration, and magnitude. Each species has its own unique signature dewlap and headbob pattern. The greatest diversity of anoles is in the Greater Antilles of the Caribbean, with as many as sixty-four species living in Cuba alone and up to fifteen species living together in a single habitat. Surrounded by all these similar but distinct neighbors, one of the biggest challenges facing an anole is to identify a mate of its own species. During courtship a male bobs its head in a species-specific series of nods while flashing its species-specific dewlap. If the female is convinced her suitor is from her own species, she may respond with her own bobs and arch her neck to indicate her receptivity.

Despite the extravagance of the male courtship display, it is probably not used by females in choosing which specific male within her species to mate with.[27] As discussed in chapter 7, many male ornaments become highly elaborated over evolution because females show a preference for mating with males that have a particular variant of the display. However, this mechanism probably does not explain the diversity of neck displays in male anoles. Males within a species vary little in their dewlaps; there are no obvious macho and wimpy variants. Moreover, when females are given a choice in the lab, they show little preference for one male over another based on his displays.[28] This is not to say that dewlap displays have no sexual function. Indeed, when a female is presented with a male that is performing the dewlap display of another species or with a male that has no display at all, she usually opts not to mate. So dewlap displays are probably more for choosing the right species rather than the singular Mr. Right. This species recognition system via the neck is evidently quite reliable, since different species apparently rarely mate with each other.

Dewlaps clearly help *maintain* distinct species in present times, but the cause-and-effect relationship between dewlap diversification and *formation* of new species in the evolutionary past is less clear. Did abundant speciation lead to great dewlap diversity, or did broad variation in dewlaps drive the formation of many new species? Evolutionary biologist Jonathan Losos argues that the unique dewlap of a species might have evolved first as an adaptation for conspicuous signaling to members of its own species and subsequently became a means to distinguish between species.[29]

Since anoles occupy a wide range of environments—dark forests, open grasslands, brown tree trunks, green leaves—they have evolved diverse patterns and colors to stand out against their backgrounds. Given this requirement that signals must be conspicuous to serve in any form of communication, Losos posits a scenario for speciation that goes something like this.[30] A species with a white dewlap living in a dark forest might extend its range into a neighboring open grassy area. In this new bright environment the dewlap of this founding population does not initially communicate effectively because it does not contrast strongly with the background. Through natural selection acting over generations, the dewlap darkens and provides more contrast with its new environment. But now that the colonizing population and the original population have diverged in their communication signals, they no longer mate and therefore become separate species. This sequence, repeated over millions of years in diverse environments, likely drove one of the greatest diversifications in the history of terrestrial vertebrates, one attributable, at least in part, to variations on a flashy throat. Nature's ornate "tattoos" at the neck consolidate membership within the species and maintain separation among species. And the world is a far more interesting place because of it.

STATUS

While signals broadcast from the neck are sometimes neutral declarations of membership ("I am a member of this group"), they frequently imply a rank ("my group is a higher rank than yours"). People sometimes display group identifiers because they want to be associated with an upper class, a more prestigious institution, or a more exclusive club. In addition, emblems at the neck commonly convey one's individual rank. As captions below our chin, our clothing and jewelry at the neck commonly serve as status symbols.

The early origins of neckties as identifiers of social class are evident in an 1820 British publication *Neckclothitania*.[31] The publication had perhaps the first mention of men's neckwear as a "tie."[32] In this gentleman's manual written for upper-class Englishmen, the author voiced his disgust about how the French and American Revolutions and their egalitarian sympathies so disrupted the "thermometer of fashion." With all the societal upheaval, it became more difficult to distinguish people by social class. With men of many backgrounds roaming around London all wearing a tie, how could you distinguish between "a gentleman and his opposite"? The author explains that you only needed to look at two features of his tie: its crisp starching and its exact knot shape. A "petty tradesman" would not be able to pay "his washerwoman's bill by having his neck-cloths starched instead of having them washed *au naturel*." The author provided explicit instructions on properly starching ties that could be passed on to the hired help. The pocket-sized manual contained instructions for fourteen distinct knots and their appropriateness for various social events. Thus a cultivated gentleman out on the town could easily carry the manual and use it to adjust his knot en route to make it fit the next social engagement. Signaling aristocratic status was all about sporting a stiff tie and tying the proper knot for the proper occasion. From this class signaling in London's streets of 1820, it's not hard to draw a line to New York's Wall Street in the 1980s. In his novel *Bonfire of the Vanities,* depicting the caste-like hierarchy in the financial sector, Tom Wolfe writes, "Pale yellow ties became the insignia of the worker bees of the business world."[33]

Like men's neckties, women's scarves have come to signify status in the corporate and professional worlds. In 2015 the BBC called the scarf the "new power symbol" for women, profiling in particular Christine Lagarde.[34] As former managing director of the International Monetary Fund and current president of the European Central Bank, Lagarde became famous for her signature use of scarves in her professional attire.[35] In earlier decades, particularly in the mid-twentieth century, scarves, like neckties, were commonly worn as markers of membership in high-society institutions. High-profile women of society such as Audrey Hepburn, Grace Kelly, Brigitte Bardot, Jackie Kennedy Onassis, and even Queen Elizabeth II commonly donned scarves on their necks and heads when out in public. A silk Hermès scarf was the ultimate status symbol. Fashion houses regularly produced limited-edition scarves that they sent as gifts to their preferred clientele.[36]

Perhaps the most extreme forms of status worn at the neck were the ruff collars of the late sixteenth- and early seventeenth-century European aristocracy. Ruffs originated from modest frills along the neckline, and within decades they evolved into separate, wildly ornate garments. By the 1580s these pleated folds of lace or fine linen, worn by both women and men, grew to "a full quarter of a yarde" in breadth and vertically reached from the chin over the shoulders. When worn by women, ruffs often covered the bust as well. The effect of this contraption was to boldly elevate and frame the head but also, out of necessity, to tilt the head backward. With all that fabric at their neck, they had no choice but to strike a "proud" posture. As cultural historian Susan Vincent writes, "Chin up, head held high, the wearer [of a ruff] assumed the carriage of haughty disdain."[37] Moreover, in displaying such a delicate, highly configured, and even restrictive ornament, ruff wearers showed that they did not need to dress practically. They were blessed with an abundance of money, time, and servants.

It is hard to overestimate the material and labor required to make a ruff collar. Vincent called ruffs "the ultimate fashion of waste"—perfect markers of wealth and opulence. The most elaborate ruffs used more than twenty yards of linen. Each inch of fabric commonly contained

Figure 18. Portrait of a woman in a ruff collar, 1644. Artist anonymous. Museum of John Paul II Collection.

over a hundred linen threads and was often embroidered with gold or silver. These ruffs required the construction of a wire frame as a supporting structure and more than a hundred pins to fasten the figure-eight folds. They took servants hours to put in place. The linen was starched and, because sweat and wear weakened the starch, the whole process had to be repeated with each wearing. The expansion of the relatively modest collar into a ruff of excessive proportions occurred over only a few generations.

In the evolution of the ruff, the key innovation leading to such exaggerated forms was the use of starch to stiffen fabric. Starching as a process in fashion began in the 1530s, just prior to the ruff craze, and, over the following decades, it allowed the ruff to grow both wider and taller. Like making ruffs, making starch itself was labor-intensive—it took over two months of soaking, washing, and drying. Moreover, starching used a commodity—usually wheat or corn—that was cheap to the aristocracy but precious to the peasantry. During periods of grain shortage, the opulence of the ruff must have been infuriating to the commoner. The grain harvests that sustained the ruffs of the aristocracy came at the expense of feeding the masses. With two pounds of grain required to make a pound of starch, it has been estimated that millions of potential loaves of bread were diverted into the fashion whims of sixteenth-century aristocracy.

At about the same time as European elites sat for portraits in their aristocratic ruffs, royalty in the West African Edo culture, in present-day Nigeria, were sculpted with their own version of status ornamentation at the neck.[38] The kings, chiefs, and other members of the court were commonly depicted in brass sculptures with up to forty rows of coral beads that extended from their lower neck up to their chin. Coral, particularly red and orange varieties, has been a prized material for more than five centuries, and it continues to have a central place in the costume culture of Benin.[39] The king, or Oba, controls the production and distribution of coral beads and so can dispense them in ways that confer status to members of his court. His own costume, of course, is the most extravagantly decorated with coral beads. In the most extreme case his ceremonial regalia includes a set of coral necklaces that stand out from his neck up to 1 foot (30 centimeters).

The Oba's wife is also adorned with many layers of coral beads covering her neck, head, and other body regions. In the next echelon of the hierarchy, the Oba's chiefs wear rings of beads that are threaded on wire and so stand up as a stiff, layered cylinder. Because the Oba uses beads in a system of patronage, the height of a chief's collar is directly proportional to his favor with the Oba and thereby his status within the court. To assure that the Oba can always know the relative rank of his chiefs, the chiefs must always wear their collars in his presence. The Edo people have a saying: "The head leads one through life's journey."[40] But the decorations just below surely signal how much clout one will carry along the way. In the Oba's court, status in the hierarchy, measured in beads, is on display at the neck for all to see.

• • •

In various emblems worn at the neck, people usually encode status through nonverbal symbols—the most exclusive scarf or necktie, the largest ruff, or the most ornate jewelry. But because these signals are codes and not words, it is easier to lie with them. Without wearing a sign that says "I am elite," we could wear an exclusive symbol—an Hermès scarf or an Ivy League tie—that would deceptively signal membership in an elite group without explicitly announcing a falsehood. In any system of communication there is always the possibility of intentional miscommunication. However, there is also the risk that if we were caught in our deception, we might incur a social cost. We might be shunned or mocked for such false implicit claims. The issue of honest and deceptive signaling at the neck was examined in a classic set of studies by Sievert Rohwer in the 1970s and 1980s. Rohwer's subjects, however, were not humans strutting around in society. Rather, they were sparrows negotiating conflict and social rank within the complex social dynamics of a flock.

Like humans, many animals use visual codes to communicate their status within a social hierarchy. It is often difficult to untangle the use of such status symbols from sexual signals, since high rank frequently confers more access to mates. However, Rohwer's studies on sparrows showed that signals displayed only in the nonbreeding season, when sexual politics were at bay, serve as badges of social status, communicating both rank and maturity. Harris's sparrows have dark plumage on their throat and crown during the winter, when they engage in frequent social conflict within the flock over resources such as food, roosting sites, and high-quality habitats—but not mates. The dark splotches correlate with dominance; the bigger and darker the badge, the higher in the pecking order.[41] It is not just for show. Individuals with darker throats tend to win aggressive contests over those with lighter throats, so their badge is an honest signal because it reliably predicts their fighting ability. Such honest status signaling reduces the overall conflict within the flock because subordinate individuals can avoid fights with higher-ranked individuals by simply glancing at their throat patch.

However, the stability of this peace could be disrupted if a genetic mutation within the population enabled some individuals with poor fighting ability to grow darker badges. Based on this misleading badge, they might rise in the hierarchy "dishonestly" and produce abundant "dishonest" offspring. Cheaters would prosper, and the whole system would fall apart. To see what would happen in this scenario, Rohwer experimentally manipulated the badges of subordinates and observed the consequences for the social system. When he increased the badge size of some subordinates by dyeing the throat feathers, thereby making an experimental cheater, the marked birds were aggressively attacked by true dominants. The flock was not deceived. They seemed to punish individuals that appeared to "dress" above their rank, and the overall level of aggression in the flock increased.

Rohwer went on to show that the importance of honesty in status signaling did not apply to all age classes.[42] In young birds high-status badges could be advantageous even when they were fake. Normally, young birds are subordinate in the flock and have small and light-colored throats and crowns. Rohwer collected a group of first-year birds and dyed the throat and crown feathers of some individuals to give them badges that resembled those of adult males. These experimental birds that had augmented badges came to dominate unmanipulated birds that retained their juvenile badges. Wearing just the right costume gave them clout even though they showed no discernible change in behavior. Evidently the need to be honest in status signaling depends on the age of the birds you flock with. Maturity seems to demand honesty.

. . .

Most signals humans display at the neck are a matter of choice. We select our collars, neckwear, and jewelry in our dressing rooms, and we select our skin art in tattoo studios. But one feature of our identity—our age—projects forth directly without, or perhaps despite, our choice. Writer and filmmaker Nora Ephron was not happy about this cruel inevitability. In her essay "I Feel Bad about My Neck," written when she was in her sixties, Ephron declared that she was actually quite satisfied with her aging appearance in general, but she confessed in regret and anger that she simply hated her neck.[43] She hated it specifically because it was honest. Anyone can conceal an aging face with make-up and injections, but "the neck is a dead give-away. Our faces are lies, and our necks are the truth. You have to cut open a redwood tree to see how old it is, but you wouldn't have to if it had a neck."[44] Like it or not, our neck reveals our demographic identity.

Ephron wrote openly and candidly about her neck, but in public appearances she always concealed her neck with turtlenecks or scarves. In private moments she sometimes pulled back the skin of her neck while standing at the mirror and gazed "wistfully at a younger version" of herself. She listed a whole taxonomy of age-related disfigurations: chicken necks, turkey gobbler necks, elephant necks, scrawny necks, fat necks, loose necks, crepey necks, banded necks, wrinkled necks, stringy necks, saggy necks, flabby necks, and mottled necks.[45] Some people are lucky enough to fit into multiple categories.

Beyond simply its visibility, the neck has several features that make it a reliable reporter of age and a marker of demographic class.[46] Over time, skin all over the body loses collagen and elastic fibers, and therefore loses its elasticity. However, the skin on the neck is particularly thin. At least compared to facial skin, neck skin attaches at fewer places on the underlying muscles. Together, these features mean that the neck skin is especially prone to sagging under the wages of gravity. Secondly, gradual bone loss from the jaw and fat deposition in the neck diminished a crisp demarcation between the jaw and the neck. Strike three: one of the superficial sheetlike muscles of the neck, the platysma, tends to split into separate cords with age, commonly yielding a ropiness to the surface of the neck.

Ephron was not at peace with such visible aging, saving her most pointed questioning for those who claim to be at peace: "Every so often I read a book about age, and whoever's writing it says it's great to be old. It's great to be wise and sage and mellow; it's great to be at the point where you understand just what matters in life. I can't stand people who say things like this. What can they be thinking? Don't they have necks? Aren't they tired of compensatory dressing? Don't they mind that 90 percent of the clothes they might otherwise buy have to be eliminated simply because of the necklines?"[47] Humans age all over our body, but it is nowhere so visible or truthful as in our necks. We are destined to go to our graves with honest badges of decline draped beneath our chin and must find a way to make peace with that inevitability. Ephron hadn't gotten there yet.

• • •

From the ascendence to presidential power to the decline of our aging bodies, the neck commonly marks people as individuals. From the excessive ruffs of the aristocracy to the indelible tattoos of street gangs, the neck also commonly marks their membership in certain groups. These neck-borne emblems range widely in how they signal identity. Some are intentional and others are unavoidable; some are situational and others are permanent; some are honest and others are fake. But they all provide a visual backdrop, a nonverbal code that modulates our social interactions. While our voices and our facial expressions are busy actors communicating the rich narratives of our thoughts and emotions, our necks and their ornaments are props below, setting the stage by conveying our social standing. The widespread use of such ornamentation across history and geography indicates they are perennial components of the human drama.

CHAPTER NINE

Power & Politics

Aggression and Control at the Neck

Edward Hicks, born in Bucks County, Pennsylvania, in 1780, was schooled in Quaker beliefs. By age thirty-three he was a traveling Quaker minister, earning money on the side as a self-taught painter. Like most Quakers, Hicks yearned for peace, and over a twenty-nine-year period (1820–1849), he painted more than sixty unique variations of the Peaceable Kingdom, a theme taken from the prophetic vision of peace in the biblical book of Isaiah: "The wolf also shall dwell with the lamb, and the leopard shall lie down with the kid.... And the cow and the bear shall feed; their young ones shall lie down together: and the lion shall eat straw like the ox.... They shall not hurt nor destroy in all my holy mountain" (Isaiah 11:6–9, King James Version). In all versions of Hicks's *Peaceable Kingdom*, the foreground highlights this zoological harmony: carnivore and herbivore, predator and prey, all at rest without fear or want. In the background Hicks often painted a human drama that portrayed some act of peace. For example, in an 1833 rendition the English Quaker William Penn and leaders of the Native Lenni Lenape tribe are signing a treaty of friendship. The idyllic imagery in these paintings is uplifting, even inspiring, and this was no doubt Hicks's intent.

Born a century later, but only fifty miles away, another famous artist, Horace Pippin, painted his own version of the Peaceable Kingdom.[1] In many ways Pippin and Hicks had similar life experiences. They both grew up in bucolic Pennsylvania, taught themselves to paint in their thirties, and were well received by the artistic establishment. But Pippin had two formative experiences that Hicks was spared. Pippin was a soldier in the trenches of World War I and returned home partially maimed. In addition, and most relevant here, he was a Black man who grew up in America during a wave of brutal lynchings. In 1911, when Pippin was age twenty-three, a mob of two thousand people lynched a Black steel worker in Coatesville, Pennsylvania, fewer than 25 kilometers (15 miles) from Pippin's hometown of West Chester.[2] So, when Pippin painted his Peaceable Kingdom, modeled explicitly on Hicks's paintings, he added a notable difference: the background included an image that alluded to the tragic vulnerability of the neck as a site of domination and oppression. Pippin's version had a noose.

Pippin's portrayal of the Peaceable Kingdom, painted near the end of his life, was a series of three canvases. He titled them *Holy Mountain*, an alternative reference to the biblical verse. In the lighted foreground, harmonious biblical animals—wolves and lambs, lions and oxen—rest together with a shepherd and children. But in the darker background you can see small disturbing images of war, and on the left side a Black man hangs by a noose. "Now my picture would not be complete of today if the little ghost-like memory did not appear in the left of the picture.... all of that we are going through now," said Pippin.[3] Nevertheless, he concluded, "there will be peace." Pippin was ready to hope and dream, but he could not completely dismiss America's history of violence and terror, particularly that exacted at the neck.

The neck is both vital and easily compressed or severed, and this unavoidable biological condition has caused it to be a locus of violence, exploitation, and control. This history of subjugation at the neck plays out in both humans and animals. Predators often attack and kill their prey at the neck. People have long slaughtered their livestock at the neck, yoked their draft animals, and used collars with leashes to control their pets. Powerful people have used simple devices—shackles, nooses, and guillotines—to restrain, terrorize, and execute the less powerful. Beast versus beast, human versus beast, and human versus human: these conflicts have complex, often disturbing histories. Many of them pivot on taking advantage of a uniquely vulnerable site, the neck. This fact dangles ominously in the dark background of human history.

BEASTS VERSUS BEASTS: PREDATION AT THE NECK

Across the tree of life, most animal species eat other animals. By one recent estimate, almost two-thirds (63 percent) of all animal species are carnivores.[4] (Herbivores are one-third of animal species, and we omnivores are very rare, comprising only about 3 percent of all species.) The ancestors of all animals as well as the ancestors of the largest animal groups (arthropods, mollusks, and chordates) were likely all carnivorous. Among vertebrates with necks, carnivorous species similarly outnumber noncarnivorous species by about two to one. Across the 800 million years of animals and the 350 million years of necked vertebrates, carnivory has been a predominant feature of life.

In the evolutionary struggle between predator and prey, necks have played a dual role. In prey they are sites of vulnerability; in predators they are adept hunting tools. A prey's neck is a primary target for predators because, to be flexible, the neck is unarmored and one of the narrowest parts of the body. Since the throat is mostly soft tissue, predators can penetrate there without the risks of teeth-to-bone collision. Moreover, because the neck transports blood, air, and neural impulses, which cannot be interrupted for even a minute, the predator can quickly subdue a large prey with a brief puncture or clamp on the throat. The limbs and head are also narrow and sometimes the site of attack, but they are bony and able to fight back with kicks or bites. The throat is soft and defenseless. By exploiting the vulnerability of their prey at the neck, predators can feed on animals with a broad range of body sizes.

Ironically, this same structure that is so vulnerable in prey is the source of great capacity in predators. Selection for a predatory lifestyle was a major force shaping the ancestral body plan of vertebrates and, in a more limited way, the neck design of tetrapods (amphibians, reptiles, birds, and mammals). The neck allows predators to scan for prey with their senses and to orient their jaws toward prey independently from movements of their whole body. That is, the neck is advantageous for both detecting and apprehending prey. Furthermore, the strength and flexibility of the neck often contribute to the capacity of the predator to deliver the final blow to its prey. Many predators use their necks to kill their prey at the neck.

Most necked predators—for instance, most amphibians and reptiles, plus insectivorous birds and mammals—eat prey that are small; they catch and swallow their meals whole. But predators that eat large prey—such as many birds (raptors), mammals (e.g., those in the order Carnivora, such as dogs, cats, weasels, and bears), and a few reptiles (crocodilians and Komodo dragons)—first subdue and kill their prey, and later rip off bite-sized pieces. Compared to predators of small prey, these predators of large prey face at least two additional challenges: where to grab a prey animal that is larger than the gape of their mouth and how to subdue their prey without themselves getting hurt. Attacking prey at the neck addresses both issues. A lion can grab a zebra by its throat, and an owl can rapidly overpower a flailing rodent by snapping its neck.

Predators kill their prey at the neck in several ways. Members of the cat family Felidae (lions, tigers, cheetahs, etc.) commonly use a throat clamp technique in which they grip their jaws around the throat high up near the jaws of the prey. There, the trachea is less girded with cartilaginous rings and can be compressed more easily by the predator's bite. Surprisingly, this form of attack is usually not very bloody; the large carotid arteries in the neck typically stay intact. The throat clamp asphyxiates the prey and also blocks any vocalizations originating from the nearby larynx, so it is largely a quiet death. After the kill, cats commonly face competition from scavengers looking for a free meal.

Because cats usually hunt solo or in small family groups, they have little ability to defend their carcasses against competitors, such as marauding packs of hyenas. Researchers have speculated that by minimizing blood loss and vocalizations from the prey, this killing technique limits the odor of fresh blood and the squeals that could draw in unwanted competitors. The prey suffocates quickly, quietly, and bloodlessly, and the predator can begin its feast with only its family and friends.

In contrast to cats, dogs (members of the family Canidae) are usually less elegant in their killing. On the whole their teeth are more fragile, and their bites are less powerful than those of cats. Their necks are longer, weaker, and less able to subdue their prey.[5] For dogs such as wolves, their strength is in their numbers and endurance. Hunting in packs, they tend to chase their large prey, such as elk or caribou, to exhaustion, with many individuals eventually slashing the prey's flank and limbs. Once the prey falters, one wolf commonly clamps its jaws around the muzzle of the prey while the pack bites into the neck. It is often a bloody affair, and the prey dies of blood loss or aspiration of blood into the lungs through the severed trachea. Because wolves hunt in a group, they can protect their kill from scavengers. It doesn't so much matter if the whole neighborhood can smell or hear the feast.

Raptors (e.g., eagles, falcons, and owls) usually catch their prey with their talons but finish them off with their bite. Eagles and hawks have tightly curved talons that are especially well designed to restrain struggling prey. With their prey tightly secured, they can take their time before dismembering their prey with their beak. However, falcons have smaller, straighter talons that grip their prey less securely, and they must subdue their prey more quickly. To do this, they have a protrusion on their beak that they use to sever the spinal cord at the neck of their victim and immediately paralyze it.[6]

In all these diverse predators the jaws are clearly important killing tools. However, in many cases the jaws are crucially aided by an accomplice just below. Predators' necks can also be vital weapons. One of history's most notorious predators, saber-toothed cats who roamed the

Americas eight hundred thousand to ten thousand years ago, had impressively long canine teeth in their upper jaw, up to 28 centimeters (11 inches) in some species, but their success as predators probably depended on their powerful necks as much as their long teeth. These cats used their thick neck muscles to help drive their sabers deep into the victim. Their famous teeth were enormously long, but they were also slender and surprisingly weak, and they probably could not withstand a lot of torsion or impact in an attack. In addition, the cats' jaws and biting muscles were not especially strong, with a biting force probably only about one-third the strength of a comparably sized contemporary lion.[7]

Given these limitations, researchers doubt that saber-toothed cats could grip their prey tightly enough to wrestle down and kill them with a jaw-driven bite to their victim's neck. The cats likely brought down prey by a surprise pounce, using primarily their limbs to overcome their victims, and then once the prey was subdued, the cats inserted their sabers into their preys' necks with a more controlled, precise, and fatal stab.[8] With such long canines, the cats could reach deep into the carotid arteries of even prey that had large-diameter necks, including bison, young mammoths, and mastodons. Because saber-toothed cats subdued their prey before biting them, their slender sabers were not endangered by flailing movements of their prey. One subsequent variation on this hypothesis is that saber-tooths might have used a "can opener" technique to supplement their relatively weak bite.[9] In this model the cats initially used their jaws to pierce the neck and anchor their lower teeth into their victim. However, the killing action came when they thrust their large canines forward and downward in an arc, like an old-fashioned ratcheting can opener, with force generated mostly from their flexing neck and limbs. Their saber teeth were only the tip of a weapon that extended down into their neck and below.[10]

Though perhaps not as dramatic as large predators hunting even larger prey, many smaller predators also use their necks for killing their prey. Their killing technique relies not so much on the strength of their neck as in its speed and flexibility. For example, owls commonly kill by grabbing the neck of their prey and swiftly twisting their own fantastically mobile neck to break the prey's spinal cord. Similarly, when small canids like foxes catch rodents and birds, they commonly thrash their head around to snap the neck of their prey. Perhaps the most extreme form of this neck-snapping technique is found in a small songbird—the loggerhead shrike.[11] Shrikes are finch-sized songbirds that routinely eat rodents, lizards, and other prey that are up to three times their body weight. Unlike raptors, shrikes do not have talons to apprehend their prey. Instead, they dive down on their prey and jab their beaks into weak points on the neck. The next details of its killing behavior happen so fast that they are only detectable with high-speed videography. After grabbing the prey's neck, shrikes shake their own heads extremely rapidly—eleven times per second—exerting a force six times that of gravity to break the cervical spine. The victim is literally shaken to death.[12]

The vital function of the neck in carnivory continues after the kill as well. Some predators transport their dead prey considerable distances to eat their meal in private or to feed their offspring, using their necks to bear the weight of the prey in their jaws. Picture a fox carrying a rabbit to its den. Shrikes, after shaking their prey to death, display another remarkable behavior that gives them their nickname, the butcherbird. They carry their prey to a high prong—a tree twig or barbed-wire fence—where they skewer their victims and store them for a later meal. Predators of larger prey, such as the "hypercarnivores" of Africa (e.g., lions, cheetahs, hyenas, and wild dogs) and the gargantuan Komodo dragons (lizards) of Indonesia, rip flesh off the carcass, using their neck to drive twisting and pulling forces to separate their fleshy meal from the bone. Crocodilians commonly dismember their prey with a "death roll," in which they grasp their prey with their jaws and rotate their whole body to twist off swallowable portions. (They have no chewing teeth to macerate their food in their mouth.) Though crocodiles cannot move as fast as shrikes, they can roll their massive bodies in a series of complete circles in less than a few seconds.[13] This whole-body death roll requires strong contractions of their neck muscles, so they do not twist off their own head. So, while some predators such as shrikes and owls actively torque their necks to kill, crocodiles feed by resisting neck torque.[14]

Carnivory is brutal. It is not pretty or fair, but it is natural and used by the majority of animal species on earth. Among terrestrial vertebrates, particularly those that eat large prey, predation is commonly a struggle of neck against neck. These predators use the versatility of their neck (its strength and flexibility) to overpower their prey's vulnerability at the neck (its narrowness and its unprotected infrastructure). This drama has played out since the very origin of the neck. And continues today. Given the central role of carnivory in animal evolution, we cannot hope for a Peaceable Kingdom in the zoological world. Nor should we. Our efforts to purge large predators—mostly for our own peace and the protection of our livestock—has only wrought major ecological imbalance. Without these predators, prey species overpopulate. This imbalance propagates through the food chain, and our ecosystems suffer. Moreover, these attempts to make the human world more peaceful by eliminating predatory violence have brought us to the brink of forever depriving the earth of some of the most magnificent creatures ever to have lived.

HUMANS VERSUS BEASTS: FOOD AND LABOR

At our origin we humans were also predators. But rather than killing with our jaws or our necks, neither of which are very strong, we commonly used our inventions, our hunting tools. Many lines of biological and archeological evidence indicate that as early as 1.5 million years ago, ancestral humans hunted extremely large prey—including elephants that were three times larger than modern elephants as well as hippos and rhinos.[15] To capture large prey, early hunters directed their weapons—spears, clubs, or rocks—to many body regions, not just the neck. But it is likely that once their large prey were brought down, they commonly used knife-like tools to cut the throats of their victims to quickly kill them and avoid injury from a protracted struggle. By as early as twelve thousand years ago, humans enlisted the help of another carnivore, dogs, to help with the hunt. These two group-hunting species, both possessing great endurance, evidently combined their different talents—human shrewdness and canine instinct—in what must have been formidable carnivorous teamwork.

Beginning about ten thousand years ago, humans developed a practice that greatly simplified the hunt for food. Our forebears domesticated herbivorous prey animals, confined them, fed them, and killed them with little chase or struggle. But even after this taming, humans usually killed domesticated prey at the neck: slitting the throats of mammals (e.g., sheep, goats, cattle) and wringing the necks of birds (chicken, quail). For many people throughout human history, killing for food was a routine, if not daily, activity. But, like many routine human activities, killing domestic animals became ritualized in some cultures. Animals were sacrificed in religious ceremonies, blessed before slaughter, and feasted on at communal gatherings. In some religions slaughter of domestic animals also came under strict religious laws: *shechita* in Judaism and *dhabihah* in Islam.[16] In both these religious traditions slaughter is accomplished by a knife cut to the neck and exsanguination through the neck's large, sliced blood vessels. Both traditions emphasize killing animals with humaneness and compassion, and both have elaborate guidelines (e.g., the kind of knife, the motion of the knife, and the training of the butcher) to minimize pain and suffering by the animal.

The scriptural basis for these laws differs between the two religions. In Judaism, *shechita* is largely about avoiding consumption of blood. According to Hebrew scripture (Deuteronomy 12:16), "you shall not eat the blood; you shall spill it on the ground like water." The *shechita* slaughtering method ensures that all the blood is quickly drained from the animal. In the Qur'an (2:173) the prohibitions are more specified and extensive. For example, "Forbidden to you (for food) are: dead meat, blood, the flesh of swine, and that on which has been invoked the name of other than Allah; that which hath been killed by strangling, or by a violent blow, or by a headlong fall, or by being gored to death." In Islam, killing by slicing the neck is the only permissible form of slaughter, and before every such killing, the butcher recites *Bismillah*, or "In the name of God." For millennia, humans have slaughtered their domestic animals at the neck, for either daily food or religious ceremony. Usually these practices are quite distinct, but in ancient Jewish and Islamic traditions and continuing in present-day Kosher and Halal guidelines, they are fused into a single act: the daily sanctification of animal sacrifice.

Most contemporary meat production is concealed behind a series of closed doors. Very few people have any direct experience with the modern "hunt"—the transformation of muscle on the hoof to meat in plastic packages. In contemporary slaughterhouses the animals are shuttled through large, sanitized facilities and stunned through various technologies (mechanical, chemical, and electrical) to render them unconscious before killing. Yet even in these high-tech corporate slaughterhouses, the animals are suspended upside down and exsanguinated from the sliced carotid arteries and jugular veins. In the neck nature has provided an effective drainage spout.

• • •

Ever since our origins, we humans have killed animals to eat their muscles. For about five thousand years we have also kept animals alive so that we can exploit their muscles for a different purpose: work.[17] After the domestication of plants and animals for food, humans began to raise large strong mammals to pull plows and carts.[18] And while it is mostly the leg muscles of these beasts that powered the pulling, the weight of the load was usually borne at their necks through yokes. Until the invention of steam and gasoline engines in the nineteenth century, humans offloaded a tremendous amount of labor onto our beasts of burden. These beasts cleared and cultivated agricultural land, they transported heavy goods to homes and markets, they hauled building materials from quarries and forests, and they pulled barges along canals. It is safe to say that human civilization was built on the backs, legs, and, yes, the necks of our draft animals. Even in our own time, animals around the world still do a significant amount of labor. In a 1979 study researchers estimated that India alone had about seventy million bullocks for pulling carts; their combined power was nearly equivalent to the country's entire electrical capacity.[19]

For food, humans have domesticated a wide variety of (mostly) mammals and birds, but for our draft animals we employ only two groups of large mammals: bovids (oxen, water buffalo, and yaks) and equids (horses, donkeys, and mules). Of these two groups, the bovids were the first used as draft animals, dating to early Bronze Age (ca. 3000 BCE) Mesopotamia, Egypt, and central Europe. Throughout history bovids have done the heaviest pulling. Compared to equids, bovids have shorter and more powerful legs. Their shorter stature and stouter, more horizontal necks make them more suitable for pulling plows and carts through yokes attached to their necks. The poles extending from the yokes of oxen to a cart pass mostly horizontally, so the pulling force is mostly the same as the direction of travel. By contrast, poles connecting a cart to the higher, more vertical neck of horses would waste much of the pulling force in the upward direction. Horses can be fitted with neck harnesses that rest relatively low and diagonal across the neck, improving the direction of pull. However, a cart attached to the top of such a collar exerts an upward choking force on the throat when the horse pulls. These considerations probably played a pivotal role in keeping early civilizations reliant nearly exclusively on bovids rather than equids for draft animals. Later, beginning in medieval Europe and China, horses were commonly fitted with harnesses consisting of a collection of straps and shafts that distributed the force low around the chest and other body regions. So, while horses have done a lot of pulling over the millennia, their necks have been mostly spared.

In addition to connecting the ox to a cart or plow, yokes commonly bind the ox to a pulling partner, another ox. Such double yokes are especially important for plowing, which often requires more power than a single ox can generate. In principle, multiple oxen could be connected individually to the plow. However, by binding them together with a yoke at the neck, the oxen must pull in unison, which is much more efficient than them pulling out of sync or in different directions. In addition, bovids are herd animals, and working with a companion tends to keep them calm and more manageable. Like in many human endeavors, teamwork in oxen improves both mood and productivity.

It is hard to conceive of all the work that beasts of burden have supplied over human history.[20] To estimate the labor-saving advantage of draft animals in agriculture, researchers observed contemporary agrarian culture in Burkina Faso and compared the agricultural output of families with and without draft animals.[21] They found that households with oxen could plow twice as much land as those without. In Ethiopia a single day of labor from a team of oxen is equivalent to four or five days of human labor. Still, these ratios greatly underestimate the full contribution of oxen, since many tasks—such as hauling logs, pulling stumps, transporting large stones—could never have been accomplished at all with human muscle alone. The morality of exploiting animals for their work is difficult to evaluate. Beasts of burden have doubtlessly suffered to allow us to increase our food production, build shelters from heavy materials, and transport our goods over distances. There are surely more and less humane ways to employ their work, but it is hard to imagine civilizations rising without the labor of these beasts borne at the neck. Perhaps at the very least, they deserve due credit.

As one of the oldest of all technologies, yokes aid the fundamental practical operations of civilization: agriculture, transportation, and trade. However, the idea of yokes has also been used in an abstract but hardly less ancient way. In many cultures the yoke has been commonly used as a metaphor, even long after it stopped being used as farm equipment. Etymologically, the word "yoke" has its origin in the Proto-Indo-European root *yeug*, which gave rise to *jugum* in Latin, *zugon* in

Greek, and *yoga* in Sanskrit. In all these derivative languages, "yoke" means the agricultural tool, but it also has a more abstract sense, meaning "union" in Sanskrit and "joining" in Greek and Latin. (The ancient Hindu practice of yoga seeks union with the Supreme Spirit.) While the early origin of the word seems to emphasize the connection between beasts, the subsequent usage of "yoke" as a metaphor usually referred to the burden and toil imposed by others through the yoke.

In Western history the yoke frequently symbolizes the wholesale domination of one people by another. In the Bible the yoke is a common metaphor for the Hebrew people's bondage and taxation during Egyptian and Babylonian captivity. In the early medieval period the English people bore the "Norman yoke" imposed by the Norman king William the Conqueror. The Bulgarians endured nearly five centuries of the "Turkish yoke" under Ottoman rule. This image of domination though "yoking" was made even more concrete in the Roman Empire. When conquering foreign lands, Roman armies ritualistically marched the defeated armies under a structure suspending a yoke as a public humiliation and admission of defeat. This ritual survives in the root of our word for "subjugate" (*sub jugum*, literally "under yoke").[22]

In surveying compilations of quotations, I found that quotable (usually famous) people applied the term "yoke" to many instances of political subjugation that extend into the modern period: Spanish domination of Mexico (Miguel Hidalgo), British domination of India (Mahatma Gandhi), and Nazi domination of Europe (Winston Churchill). Throughout history quoted people (e.g., Aeschylus, William Lloyd Garrison, Jamaica Kincaid) have also referred to slavery, the most brutal form of subjugation, as a yoke. Other authors have applied "yoke" to broader systems of domination and exploitation or their abstract ideologies: oppression (Martin Luther King, Abraham Lincoln, James Madison), poverty (John Kennedy), capitalism and fascism (Joseph Stalin), and gender inequality (Abby May Alcott). In these examples of metaphorical yokes, the words reference a collective experience, the oppression of a group of people. But some authors have also likened yokes to impulses in our individual inner lives that confine or burden us: attachments (Buddha), certainty in our own opinions (Ralph Waldo Emerson), and "the yoke of our own wrong-doing" (George Eliot).

Although not a yoke per se, one of the most peculiar but persistent images of psychological burden is an "albatross around the neck" from Samuel Taylor Coleridge's *The Rime of the Ancient Mariner.* In Coleridge's poem the narrator recounts his travels and travails aboard a sailing vessel that is diverted by a storm into the remote and frigid waters of Antarctica. The ship is guided out of the ice jam by an albatross, who brings with it a wind to set the ship back on course. In Coleridge's time many sailors viewed albatrosses as good omens with supernatural spirits.[23] This was based in part on their ability to glide endlessly above the waters; they embodied the lost and wandering souls of those at sea. In *The Rime of the Ancient Mariner,* the albatross perched on the ship's rigging "like a Christian soul," and it was hailed by the crew as a savior. But one fateful day the mariner shot the albatross with his crossbow for no apparent motive. Without their avian protector, they lost the breeze, and the ship was stranded in the doldrums. The crew members grew so parched they could not speak. To punish and humiliate their murderous mate, the sailors fastened the dead albatross around the neck of the mariner. All the crew died, and while the albatross eventually fell from the mariner's neck after he gave a blessing to other sea creatures, he never outlived the psychic burden of this "hellish thing." And the odd metaphor continues to live in Western culture when we speak of a psychic load as an albatross around our necks.

Among these images of burden and exploitation, it is also worth recognizing that the metaphorical yoke sometimes has a more joyful, life-giving meaning. This image of the yoke highlights the original meaning of "to join" and conveys a sense of partnership and lasting connection to another. In these often-underplayed connotations, the yoke gives people companionship and direction, binding humans to each other and to higher ideals. For some writers love is a yoke—"a peremptory hunger," in the words of George Eliot—that ties us together in the most intimate, irresistible, and beautiful sense. This kind of bondage is mutual and sustaining. Maggie Tulliver, a character in George Eliot's *The Mill on the Floss*, calls the yoke of love a "wonderful subduer." Rupert Birkin, in D.H. Lawrence's *Women in Love*, grows to see a new "way of freedom . . . which accepted the obligation of the permanent connection with others, and with the other, submits to the yoke and leash of love." These characters came to view the yoke of human relationships as one that gives devotion and direction rather than burden and subjugation.

The concept of the yoke seems to have evolved in this direction within ancient Judaism and early Christianity.[24] In earliest Hebrew texts the yoke symbolized control and servitude within human hierarchies: the owner and the owned, the master and the enslaved, the ruler and the captive. Later, the concept of the yoke was applied to the relationship between God and people.[25] But rather than stressing the toil and subjugation of people, this notion of the yoke emphasized the sovereignty of God. In yoking to God, people lived in a close, enduring relationship with God. The Hebrew writers came to "view bearing the yoke of God's sovereignty as joy, honor and privilege rather than tragedy, hardship and sorrow."[26] Similarly, in the New Testament Jesus offers his yoke to the burdened to make life easier. "Come to me, all who labor and are heavy laden, and I will give you rest. Take my yoke upon you, and learn from me; for I am gentle and lowly in heart, and you will find rest for your souls. For my yoke is easy, and my burden is light" (Matthew 11:28–30, Revised Standard Version). Here, the yoke involves a liberating commitment, taking on a spiritual discipline that guides the follower in the "the Way."[27]

At a material level the yoke is an extremely practical tool. For millennia it allowed humans to ease their own burdens by offloading their labor onto the necks of larger and stronger beasts, usually bound together as well as hitched to their load. The three relationships surrounding the yoke—between the master and the beasts, between the paired beasts, and between the beasts and the load—encapsulate many of the archetypal relationships in life and labor, from power and dominance to partnership and teamwork. It is little wonder, then, that the yoke has permeated our literary and religious attempts to make sense of human existence.

HUMANS VERSUS HUMANS: EXECUTION AND TERROR

Like the yoke, the hangman's noose originated deep in human history with a specific function and later became a potent symbol woven into many strands of human culture. "The noose has served as a weapon, spectacle, ritual, artifact, relic, symbol of ultimate state justice, a symbol of ignominy, as a way of defining victims as inferiors," writes Jack Shuler, author of *The Thirteenth Turn: A History of the Noose.*[28] Shuler chronicles the history of the noose as a simple tool used by powerful people—members of formal legal systems or extralegal mobs—to execute individual criminals and as a weapon to terrorize entire groups of oppressed peoples with psychological intimidation. In colonial America and the United States alone, approximately fifteen thousand people have been killed by hanging in our four-hundred-year history; nearly one-third of these deaths were meted out outside the law.[29] And this is just a small fraction of the hanging deaths that have occurred in global history. The story of the noose is horrific, measured both in numbers and in suffering.

As an instrument of death, a noose takes advantage of any of three vital conduits in the neck: the trachea, the large blood vessels, and the spinal cord.[30] In all hangings, the victim has a rope with a collapsing hangman's knot placed around the neck and then is dropped or suspended from some high structure—a tree branch, a telephone pole, or a gallows. The precise mechanism of death depends on the details of the execution method. In the most calculated formal executions, the victim falls through a trap door with enough force to break the neck at the second cervical vertebra and tear the spinal cord: a quick death. But in times and places with less humane hangmen or cruder gallows—for example, in lynchings and most premodern executions—the drop was not sufficient to sever the spine. Instead, the tightened noose caused prolonged suffering by blocking airflow to the lungs or constricting blood flow to the brain. The victim commonly remained conscious for minutes after the drop. Even when the large, more superficial carotid arteries are occluded by the noose, enough blood can pass through the arteries traveling within the vertebrae to keep the hanged person alive for a brief period. Although centuries of practice and technology improved the efficiency of hanging, it was usually a gruesome, painful death.

Hanging is an ancient but not prehistoric form of killing. Humans executed each other in many other ways long before the noose. The first direct evidence of a hanging execution dates to a corpse from the fourth century BCE found in a Danish bog, whom archeologists named "Tollund man."[31] "Tollund man" clearly died with a rope encircling his neck. Because he was buried in the anoxic conditions of a bog, his physical features are extremely well preserved, including his detailed facial features and the individual strands on the woven leather noose. Due to this fine-grained preservation, archeologists can see that the man did not apparently endure a death of abuse and disgrace as was typical of other criminal executions. His face appeared calm, his body was unharmed, and his burial seemed respectful. This direct evidence, along with depictions in contemporaneous tapestries, indicate that "Tollund man" was likely hanged as a ritual sacrifice to the Norse god Odin. As homicides go, it was probably a relatively honorable death.

Hanging was also documented in the written records of ancient Greece from the same approximate historical period. Usually, hanging shows up as a disgraced form of suicide, rather than an execution. For example, in Sophocles's play *Oedipus Rex*, Jocasta hangs herself in shame upon discovering her accidental incest with her son/husband Oedipus. Four hundred years later in the Roman world, Judas Iscariot hanged himself in penitence after seeing that his betrayal led to Jesus's condemnation and crucifixion. While Jesus's crucifixion was a public act of the Roman state, Judas's hanging was a solitary act of self-punishment.

These early depictions suggest that the noose, which later came to represent ultimate power of the state or the mob, may have originated for forms of death other than execution.[32]

After crucifixion was banned in the fourth century CE, hanging became a frequent means of capital punishment. In the medieval period, gallows were commonly erected on prominent hills or at crossroads, where the hanging and putrefying corpses were in full public display. This protracted and public form of death was exacted primarily on the lower classes; they did not deserve a quick, honorable beheading. As Schuler put it: "The mode of punishment was important because it telegraphed to the onlookers the rank or class of the victim—the confined nature of the slow strangulation of the hanging death (and the ignominy of the rotting corpse) was less desirable than the quick cut of the blade." Over time, the lower classes increasingly faced the gallows as the authorities expanded the range of capital crimes from violent offenses (e.g., murder and rape) to property offenses that were most often committed by (or attributed to) the poor (e.g., theft and counterfeiting). By the eighteenth century, English law codified a list of more than two hundred crimes that were punishable by the noose.

While execution by hanging became more common in Europe, so did an exemption. When clerics were accused of a capital offense, they could evade the noose by claiming the "Benefit of the Clergy." This exception originated in the twelfth century in the movements following Thomas Becket, who as Archbishop of Canterbury resisted King Henry II's encroachment on the independence of the Church. In the aftermath of Becket's resistance, clerics could petition for their trial to be transferred from lay court to ecclesiastical court, and this usually meant they were spared capital punishment. Soon, however, the courts had to contend with the problem of how to identify a person as a legitimate cleric, since of course many common criminals claimed to be clergy in hopes of receiving the Benefit.

By the mid-fourteenth century the criterion for recognizing a cleric became codified: if a person could read Latin, it must mean they were educated by the Church into the priesthood. In trials the accused were presented with a text, and if they could read it before the judge, they were granted clerical immunity. This text became known as the "neck verse," since its proper recitation could save a person from the noose.[33] Initially the specific texts varied, but eventually the courts landed on a standard text, Psalm 51, which begins: "Have mercy upon me, O God, according to thy lovingkindness: according unto the multitude of thy tender mercies blot out my transgressions" (King James Version). Still, there were rampant cases in which laypersons memorized the verse or were coached through this test to save their lives. Over the centuries the neck verse lost its pretense as a criterion of membership in the clergy and became little more than a literacy test and a way to exempt the privileged, educated classes from execution. After a few hundred years as a loophole in the law, the neck verse was abolished by Queen Anne in 1706.

Across the Atlantic in the early American colonies, clergy played a central role in hangings.[34] The noose was used primarily to punish crimes of morality—for example, blasphemy, idolatry, sodomy, and bestiality. In the hanging ceremonies, the clergy gave sermons at the gallows to offer the crowd lessons in the moral and theological significance of the event. They directly addressed the condemned to elicit a repentance or confession. Through the eighteenth century, hangings became more common and increasingly used in convictions of property crimes. Public attendance swelled to impressive sizes. In New London, Connecticut, in 1753 a woman was hanged in front of approximately ten thousand people for allegedly killing her own baby. In Providence, Rhode Island, in 1774 a rapist was hanged before more than twelve thousand onlookers.[35] For a period, hangings were the largest of all public events, and it is estimated that most colonial Americans of this period witnessed at least one such execution in their lifetime.

Hangings in New England ended by 1835, but they continued in the southern United States. Here, too, they were mass spectacles, sometimes even with specially arranged trains to bring in spectators from far away. Enslavers commonly forced their enslaved laborers to attend the hangings when the victims were enslaved people to crush any ideas of "willfulness" or liberation. Such racist hangings were not limited to the South. In 1862, in Mankato, Minnesota, thirty-eight Dakota Indians were publicly hanged for their uprising against a corrupt federal "Indian System" that oversaw treaty agreements and trading policies on the Dakota reservation. On December 26 the Dakota prisoners were led onto a massive gallows platform and fitted with nooses. The hangman cut a single rope, collapsing the support scaffolding under the men. All thirty-eight dropped at the same time to their death. It was the largest simultaneous execution in US history.[36]

After the Civil War most hangings were lynchings—that is, executions by mobs acting outside the judicial process. While most people hung before the war had been convicted white criminals, the vast majority after the war were nonwhite victims of lynchings. Of the approximately forty-seven hundred lynchings in the late nineteenth and early twentieth century, about three-quarters were Black, and most of the remaining were Mexican American or Native American. These killings were usually done in the darkness of night, but the corpses were left hanging through the day to warn minorities of the lethal power of the white majority mob. Lynchings subsided after the 1930s, but the threatening display of the noose continued. The noose "became a stand-in for vigilantism, for murder by community, an unveiled threat and a symbol to brandish to keep blacks, especially, 'in their place,'" writes Shuler.[37]

In contemporary times the function of the noose as an instrument of governmental execution has almost completely disappeared—the last state-sanctioned hanging in the United States was in 1996, and only a few nations worldwide (and on rare occasions) currently execute with a noose. But the noose as an instrument of intimidation is still with us.[38] From 2010 to 2014, Shuler recorded almost a hundred incidents in the United States in which a noose was used in an act of racial intimidation.[39] The *Washington Post* reported that from 2015 to 2021, fifty-five nooses were found hanging in construction sites alone, including seven noose incidents in a single month (May 2021) at an Amazon construction site in Windsor, Connecticut.[40] Even more recently (November 2022), work at the project site of the Obama Presidential Center in Chicago was halted when builders discovered a hanging noose.[41] The executioner is gone, but the noose continues to terrorize. "Fear is a noose that binds until it strangles," wrote African American poet Jean Toomer.[42]

With a brutal history of racial lynchings and terrorism that continued through the twentieth century, the American psyche carries particularly deep scars from our history with the noose.[43] But noose imagery extends overseas as well, making the headlines recently in New Zealand. In 2021 a Māori politician, Rawiri Waititi, came to the Parliament floor without the required necktie, which he referred to as a "colonial noose." In its place Waititi wore a traditional Māori carved stone pendant, saying that it was "Māori business attire." He was ejected from the chamber for violation of the dress code. Later, Waititi wrote in an opinion piece, "I took off the colonial tie as a sign that it continues to colonize, to choke and to suppress [the Māori]."[44] One man's formality is another man's noose. Within days, Trevor Mallard, the speaker of the New Zealand House, reversed course and lifted the requirement for ties.

• • •

The neck is arguably one of the most political of all body parts. It generates the voice, the tool of political persuasion. Power in a democracy, at least in theory, depends on words arising from throats rather than swords pulled from sheaths. Speeches delivered to crowds and votes cast in deliberative chambers have all tilted balances of power. But the neck is also the site of political domination. Tyrants shackle their political prisoners and behead their political opponents at the neck. Political expression rises and dies just below our chin. From this side of history it is hard to believe: the instrument that would eventually sever the necks of tens of thousands of people, the guillotine, was deployed at the height of the Enlightenment and in the spirit of equality.[45]

Before the French Revolution most executions occurred through two very different means. As described earlier in this chapter, hanging was usually slow, excruciating, and ignoble, and used for the peasantry and the most disgraced criminals. Beheading was a faster, less painful, and more honorable death, and was reserved for upper-class criminals convicted of capital offenses. This two-tiered practice struck some eighteenth-century Enlightenment thinkers as unfair as well as inhumane. Everyone, regardless of caste, deserved an equally painless and dignified death. In October 1789, only two months after the newly formed French National Assembly passed the Declaration of the Rights of Man and of the Citizen, a prominent physician and assemblyman proposed a bill to address the inequities in the penal system. Article 6 of the proposal from Joseph-Ignace Guillotin read: "The method of punishment shall be the same for all persons on whom the law shall pronounce a sentence of death, whatever the crime of which they are guilty. The criminal shall be decapitated."[46] The bill was voted into law two years later as a bold act of equality and human rights.

Guillotin was looking for a more reliably humane form of decapitation. Beneath the lofty debates of human rights and penal reform, there was also a practical matter. The chief executioner, Charles-Henri Sanson, argued that it was simply unfeasible to dispense with the backlog of capital cases with a sword. They needed a more efficient killing machine.[47] The prototype for such a tool was designed by surgeon Antoine Louis, and for a brief period this new falling-blade device was called the "louisette." However, much to the dismay of Guillotin, the device was soon named the "guillotine." Guillotin and many of his allies were mostly opposed to capital punishment but saw the new machine as an improvement to the process until executions could ultimately be phased out. In the meantime, the guillotine would "deliver pristine justice, one rolling head at a time," writes author Edward White.[48]

Figure 19. The execution of King Louis XVI by guillotine during the French Revolution, 1793. Artist anonymous. Musée Carnavalet, Histoire de Paris.

After the initial demonstration of the guillotine in April 1792, the heads began to roll with little delay and with remarkable frequency, including those of King Louis XVI, Marie Antoinette, and a parade of other aristocrats and clergy of the ancien régime.[49] Even those from the highest echelons of society received the same fate as the lowly criminal: *égalité* at the basest level. The public initially reacted with enthusiasm.[50] Sanson, the chief executioner, became known as the "Avenger of the People" and his flamboyant uniform was incorporated into men's streetwear. Women sometimes wore replicas of the guillotine as jewelry dangling from their ears or as brooches pinned to their outfits.

The next chapter of the guillotine is well known. As paranoia spread through the Revolution, egalitarians turned into tyrants, and "pristine justice" turned into a bloody massacre that Guillotin would never have predicted. Under Maximilien Robespierre and the Reign of Terror, more than sixteen thousand people deemed enemies of the Revolution were guillotined between June 1793 and July 1794. Although the guillotine was used occasionally in France until 1972, the period of mass killing was brief, concluding largely with the execution of Robespierre himself in 1794—by guillotine. Ever since, the image of this machine designed to efficiently sever the neck persists with disturbing irony alongside "Liberté, égalité, fraternité" as a legacy of the French Revolution.

• • •

The US penal system never executed anyone via guillotine and has not used the noose for almost thirty years. Yet one of the most potent and influential images of the twenty-first century has been one of political violence at the neck. On May 25, 2020, George Floyd, a Black man, was killed by Derek Chauvin, a white police officer, who kneeled on Floyd's neck in a Minneapolis street during an arrest for allegedly passing a counterfeit twenty-dollar bill.[51] With Chauvin's knee on his neck, Floyd said over twenty times that he could not breathe. He pleaded for his mother. Six minutes later, Floyd became quiet and motionless, and the accompanying officers could not find a pulse. Chauvin kept his knee in place for three additional minutes. Still unresponsive, Floyd was taken by ambulance to the hospital, where he was pronounced dead after an hour. Much of the episode was captured on video by a bystander, and the footage went viral across the world. By one estimate the video and other related clips were viewed more than 1.4 billion times in the subsequent twelve days. It has become one of the most infamous nine minutes of the twenty-first century.[52] The public responded immediately and massively. Within days, protests erupted in 140 US cities and in 60 nations around the world. The Black Lives Matter protests reached their peak two weeks later with demonstrations in over five hundred US locations attended by about twenty million Americans, making it perhaps the largest protest movement in US history.[53]

But why was there such an unprecedented outpouring over this particular death? In the previous decades, hundreds of unarmed Black people had been killed by the police, mostly by the guns of white officers. Scholars and journalists have written volumes about this question, and it doubtlessly has many answers. I wonder if the mechanics and anatomy of Floyd's death played an important role. The masses could relate all too well. Although few people have experienced being shot, we all have experienced the loss of breath and the sense of urgency in restoring it. Nine minutes of suffocation is just imaginable enough to be horrifying. Many people have experienced being confined, but the idea of being compressed at our most vulnerable anatomy by the weight of another is almost enough to make us panic.

In addition to these visceral, instinctive reactions, many people surely responded to the cultural symbolism. The episode had many of the optics of a racial hanging—prolonged suffocation of a defenseless Black man by a group of powerful (mostly) white men. The image called forth ancient gestures of subjugation—for example, Roman imperial soldiers standing on the necks of the conquered. Decades before the George Floyd incident, Malcolm X alluded to this ancient image when he spoke against the burden of contemporary racial subjugation: "That's not a chip on my shoulder. That's your foot on my neck." In the aftermath of George Floyd's death, President Joe Biden extended the metaphor to our collective American experience in his 2020 address to the US Congress. "We have all seen the knee of injustice on the neck of Black Americans."[54] The tragedy of George Floyd was anatomically specified.

• • •

The history of violence at the neck is severe and ancient, extending back to the earliest necked animals. For large carnivores it is an effective means of subduing their prey. For most of human history people have killed animals at the neck for food and yoked them for labor. We have also directed such violence at other humans, executing criminals, opponents, and subordinates, all at the neck. But there are signs that the use of power directed at the neck is diminishing in human societies. In almost all places we use machines rather than draft animals to pull and plow. We have legislated protections for the humane treatment and slaughter of livestock. We punish criminals without nooses, and in many places we have criminalized the display of a noose as an act of hate. We settle political conflict without guillotines, and we protest by the millions in response to police homicide at the neck. In just a few generations, with our voices and our votes, we have curtailed the long human history of violence at the neck.

In his art Horace Pippin confronted human violence in its most brutal forms. He painted many harrowing images of war and slavery. He painted abolitionist John Brown going to the gallows. However, when he painted his *Holy Mountain* near the end of his life, he placed the noose in the dark background and a vision of peace in the foreground. Human brutality and subjugation are undeniable, but for Pippin those cannot fully obscure the ideal. As he explained: "*Holy Mountain* came to my mind because the whole world is in such trouble, and in reading the Bible (Book of Isaiah XI:6) it says that there will be peace in the land. If a man knows nothing but hard times he will paint them, for he must be true to himself, but even that man may have a dream, an ideal—and *Holy Mountain* is my answer."[55]

We all search for our own answers to the troubles of our times. Like Edward Hicks and Horace Pippin, we all yearn for peace. In doing so, we also seek a reckoning with the nooses of our times as well as the yokes of our lives. Both collectively and individually, we confront the neck as a locus of domination and burden.

CHAPTER TEN

Shields & Saints

Protection and Healing at the Neck

Most of us, at least on occasion, have felt threatened or beleaguered at the neck. Our sense of that vulnerability moves sporadically between several levels of awareness—from unconscious instincts to subconscious undercurrents to conscious imagination. But equally, we possess a range of counteractive responses to protect and soothe our necks. Such defensive measures, rooted deep in our biological and cultural past, also originate at different levels of awareness. When the body is invaded by pathogens through the mouth and nose, we unconsciously launch an immune defense, concentrated at the neck. When we perceive an external threat, aimed either directly at our necks or just generally toward our well-being, we subconsciously tighten our neck muscles. When we venture into the cold, we intentionally reach for a scarf to insulate our necks, and when we engage in dangerous activities such as warfare, we invent armor to shield them. When protection seems utterly beyond our capacity, we turn to divine intercession or supernatural powers to guard and heal our necks. From protective impulses that arise unconsciously to those that arise from our most inventive and imaginative thinking, humans resist the diverse threats the world directs at our necks.

Natural selection has been no less creative, endowing animals with an ingenious array of defensive neck structures and behaviors. Because many carnivores earn their meals by taking advantage of the frailty in their victim's neck, all groups of terrestrial vertebrates have evolved some sorts of neck defenses. Among these, reptiles seem to have evolved a particularly diverse and innovative range of antipredator adaptations centered at the neck. Certain reptiles deter predators by concentrating poisons at the neck, some flash warning or deceptive signals at the neck, and some use their necks to completely hide the head. In certain cases these adaptations guard against threats aimed particularly at the neck; in other cases they are responses by the neck to protect against general threats to the body. The neck can be the protector as well as the protected. Although some terrestrial vertebrates use their necks in active antipredator behavior, all terrestrial vertebrates use their necks in another universal but unconscious struggle: defense against the stealth invasion of microscopic pathogens.

INTERNAL DEFENSE: IMMUNITY AND LYMPH NODES

In humans as in other terrestrial vertebrates, the skin is largely impermeable to disease-causing bacteria and viruses. Most such pathogens enter the body through the thin, moist surfaces of the nose and the mouth. Our anatomy reflects this particular threat from invasion at the head: we deploy our most concentrated immune defenses in our necks, just below these entry points. The neck is densely scattered with pea-sized lymph nodes that house white blood cells whose job is to monitor and attack microscopic intruders. Of the eight hundred or so lymph nodes in the human body, nearly half are located in the short, narrow segment between the head and chest. During evolution lymph nodes originated in reptiles, the first vertebrates with an extended neck, and during embryonic development the cervical nodes are the first of all lymph nodes to form.

Lymph nodes are part of a large network, the lymphatic system, that has an intimate and unfiltered connection to the immediate environment of the body's cells. The thin vessels of the lymphatic system permeate intercellular spaces, and because they have large pores, they can collect large bacterial cells, viral particles, and cellular debris present in the fluid surrounding the cells. By contrast, the capillaries of the circulatory system can absorb only small molecules through their relatively small pores, and these small molecules do not signal much about pathogen invasions or tissue damage. So the lymphatics provide the passageways into the battle trenches in our warfare against microbes. Fluid collected by the lymphatics, lymph, is continuously transported short distances from the pathogen's entry points in the nose and mouth to the lymph nodes. Given the density of cervical lymph nodes, this distance is only a few millimeters to centimeters. These lymph nodes contain a garrison of diverse white blood cells, and each cell type has a specialized defensive role. As "captured enemies" are carried by the lymph into the node, immune cells termed macrophages recognize and engulf the pathogen. Macrophages and other pathogen-consuming cells snip out certain signature molecules from the pathogen called antigens and present them on their cell membranes. These antigens announce the pathogen's invasion to another set of white blood cells, lymphocytes. After detecting these foreign antigens, activated lymphocytes divide to make more lymphocytes that can make additional direct attacks on incoming pathogens. Still other lymphocytes make antibodies to impede the pathogens' survival or mark them for destruction.

While macrophages remain in the nodes, dendritic cells prowl among the tissues, and when they capture pathogens, they migrate to nearby lymph nodes and, like macrophages, present these new antigens to lymphocytes to further amplify the immune responses. As the local battle against infection wages on within and near the lymph node, some white blood cells initiate defensive responses all over the body by secreting chemical messages into the blood that activate immune cells in distant tissues. These local warriors may also migrate out of the lymph node through the lymphatics or bloodstream to further fight distant infections. Because the lymph nodes of the neck occupy a front-line position near vulnerable sites of entry in the oral and nasal cavity, they serve as crucial citadels in our ongoing defense against pathogens. Yet, except on occasion when the deployment of immune defenders is large enough to swell our cervical lymph nodes, all this elaborate cellular warfare occurs beneath our awareness.

Our immune system guards us against domestic enemies as well as foreign invaders. Cancer cells arising within the body are also targeted by white blood cells. We all produce tiny tumors throughout our lives, but in most cases these cancerous cells are quickly vanquished by lymphocytes. Cancerous cells express molecules on their surfaces that identify them as "domestic terrorists" and that activate several classes of lymphocytes, including the ominously named natural killer cells. Rather than engulfing the enemy, these cells secrete signals that trigger a series of reactions within the cancer cell, the programmed death pathway, leading to its "suicide." New immunotherapies to treat head and neck cancers seek to enhance this lymphocyte-activated tumor suicide.

In instances when cancer cells evade detection or attack by the immune system, they may enter a mobile, more aggressive phase, known as metastasis, in which they colonize other regions of the body. While some forms of cancer metastasize through the bloodstream, cancers of the head, neck, and thyroid usually travel through the lymphatics. The lymphatic vessels, which normally transport our immune defenders, become avenues for subversive cancer to infiltrate the body. Thus treatment of head and neck cancers commonly involves targeted removal of the cervical lymph nodes and their connecting lymphatic vessels. The varied defenses the human body mounts are so complex they fill countless volumes of medical texts and scientific journals, yet any immunologist would readily admit that we have only a superficial understanding of all these defenses entail. Every year, scientists discover new actors on both the defense and offense. Our survival depends on these defenses, and our necks, positioned on the frontlines near the common points of invasion, play an especially important role. Yet, even with such high stakes and complex drama, these defenses work beneath our consciousness.

MUSCULAR DEFENSE AND HEALING TOUCH: TENSION AND RELIEF

While humans have always faced threats from pathogens and cancer, many contemporary health problems arise in response to the perceived threats of modern society. We are increasingly stressed and lonely, and our bodies often react subconsciously with a series of physiological responses, including muscular tension in the neck. These reactions probably had great survival value in our evolutionary past when we more commonly faced threats from predators and other sources of physical injury. Even today, contracting neck muscles to stabilize the head may protect us in emergency situations—falling off a bike, bracing for a car collision, or riding a roller coaster. But prolonged tension arising from chronic psychological stressors usually just makes us miserable. This protective impulse literally becomes a pain in the neck.

The familiar effect of psychological stressors on neck tension was demonstrated experimentally through a series of studies examining the contraction of neck muscles in response to emotional and cognitive challenges. In one study, subjects were given a fast-paced color-word test in which they had to quickly identify mismatches between the meaning and color of a word (e.g., the word "blue" written in the color red).[1] In a second test, subjects had to solve arithmetic problems while an overseer continuously notified the subjects of their errors. In a third study, subjects performed a simple physical task—holding their arms forward at a 45-degree angle for one minute—that required muscular contraction but was not psychologically taxing. Using electromyography (EMG), researchers recorded the electrical activity in the trapezius muscle, the long triangular muscle spanning from the neck and upper back vertebrae to the shoulder blade and forming the muscular "wings" at the base of the neck. (This is often the first muscle rubbed in a casual neck massage.) Researchers also took measures of physiological stress, such as blood pressure and levels of stress hormones. They found that the two psychological tests increased electrical activation of the trapezius and that the magnitude of this activation correlated positively with measures of physiological stress. Moreover, performing the psychologically stressful tasks simultaneously with the physical task amplified the electrical activity elicited by just the physical task. That is, stress alone tightens the neck and also compounds the muscular tension arising from light activities that would otherwise not cause discomfort.

But does this muscular response to psychological stressors occur throughout the body, or is it specific to the neck? To address this question, researchers in later studies took EMG measurements in the trapezius muscle and five additional muscles in the hands, arms, and legs while subjects were given two rounds of the same color-word stressor test.[2] They also quantified the subject's level of anxiety through a short questionnaire. During the first round of stressors, EMG activity in all six muscles increased along with self-reported anxiety. So it appears that the whole body tenses as an initial response to psychological duress. However, when the same stressor was repeated three minutes later, the results were quite different. The EMG activity in the hand, arm, and leg muscles all decreased, but the activity in the neck muscle (the trapezius) remained high. This protracted tension in this neck muscle occurred even while the subjects' anxiety decreased and their scores on the cognitive tests increased. Thus, unlike other muscles, our neck muscles are apparently reluctant to adapt even when our conscious perceptions and objective performance tell us circumstances have improved. We hold memories of the day's accumulated stressors in the neck, contracting its muscles with prolonged and maladaptive tension.

Even when we are not stressed, our twenty-first-century lifestyle commonly puts us into postures that strain our necks and give us discomfort. The neck's anatomy works best when the head is mobile and balanced, yet many of us spend large parts of our days with our heads fixed forward for hours as we work at computers and other machines or with our heads tilted down to operate our phones. When the head leans forward, it weighs two to six times more than when it is upright, and we must exert an equal opposing muscular force to prevent the head from folding further down.

To support the head at this intermediate angle, muscles in the back of the neck generate force even while they are elongated. This so-called eccentric contraction has certain energetic advantages, but it tends to cause microlesions and cellular damage in the muscle, leading to soreness, stiffness, and swelling. In addition, the forward head posture disrupts the healthy curvature of the cervical spine. The spine typically curves a bit forward (toward the throat) at the base of the neck and curves backward (away from the throat) as it rises up toward the skull. Like the design of a stone arch, this curvature helps distribute the forces across the neck vertebrae and stabilize the whole structure. But when the head tilts forward, the cervical spine straightens, stacking the vertebrae in a slanted column. Over time, the loss of the normal neck curve increases the compressive and slippage forces on the vertebrae and requires additional baseline muscular tension to stabilize the head.

Some of the problems we experience at the neck originate from postural problems well below the neck. "Our neck is the ultimate compensation organ," says Scott Raymond, a massage therapist and teacher of kinesiology.[3] It commonly works overtime to correct issues arising from our stooped thorax and sedentary lifestyles. During long periods of sitting, we tend to hunch our shoulders forward and collapse our chest. In this slumped position, the diaphragm and breathing muscles attached to our ribs become compromised in their ability to expand the thorax, and sometimes we compensate by enlisting secondary respiratory muscles that extend into the neck. One such set of muscles, the scalenes, connect the cervical vertebrae to the upper two ribs and, in compensatory breathing, lift these ribs for inhalation. Their chronic overuse as accessory breathing muscles can exacerbate tension in the neck. Moreover, because several important nerve tracts of the neck and arm (the brachial plexus) pass between the scalenes, prolonged tension in these muscles can cause pain by directly compressing these nerves.

A second set of short neck muscles at the base of the skull, the suboccipital muscles, are commonly recruited to readjust our head position when we have hunched posture. Raymond believes that our neck musculature and its nervous control are fundamentally designed to keep our gaze on the flat horizon. So, if our cervical spine tilts forward in a slumped posture, we reflexively contract the suboccipital muscles that pull the skull back to keep our head and gaze pointed forward. You can feel this reflexive connection between vision and the suboccipital muscles through a simple experiment on your own body. With your head oriented straight forward and eyes closed, press your thumbs deeply into the musculature at the back of your neck, just below the base of the skull. Now move your eyes up and down. You will feel the suboccipital muscles contract involuntarily in response. So, when the eyes are not level, the neck muscles automatically attempt to realign the head with the horizon. This compensatory tightening of the suboccipitals can directly cause muscular pain. In addition, these muscles have recently been implicated in another kind of pain: headache. Anatomical studies in just the past few decades showed that the suboccipital muscles have processes that extend inward to the outermost wrapping of the spinal cord, the dura mater.[4] Muscular forces transmitted through these so-called myodural bridges normally serve to protect the spinal cord when the neck extends backward. However, the dura mater is very pain sensitive, and prolonged tension through the myodural bridges likely contributes to stress-induced headaches.

While such stress-induced headaches are difficult to ignore, many forms of neck pain seem to move periodically between our subconscious and conscious minds. Tension in the neck commonly accumulates throughout the day, but we are often not aware of it until we take a break later in the day. Usually, a soft pillow or light movement or ibuprofen is enough to relieve the day's tension. But sometimes we turn to the caring hands of others to loosen the knots or contortions of our lives borne at the neck. Sometimes these are the hands of lovers or friends; sometimes they are the hands of professionals.

Because of the neck's distinct anatomy, neck massage requires a unique approach. When massage therapists work in most other body regions, they are manipulating relatively large muscles that lie in uncluttered anatomical neighborhoods. In these regions therapists can push and knead with their whole hands with little concern for pinching or damaging nearby tissues. But working on the neck, especially in the side and front, is different. The muscles are smaller, sometimes spanning only a few inches, and they are surrounded by numerous delicate tissues: superficial blood vessels, interwoven nerve tracks, soft glands, and lymph nodes. Neck massage is transmitted more through sensitive fingertips than strong hands. Because the neck is so sensitive and vulnerable, most patients will only relax in response to touch if they have an especially high degree of trust in their therapist.

When therapists press their fingertips into spaces of the neck, they probe the texture, mobility, and temperature of the underlying tissue. Softness and elasticity in the muscle indicate it can fully lengthen when given unconscious commands from the nervous system; hardness and stiffness mean it is stuck in a tonic contraction and has little capacity to lengthen or shorten. Puffiness might come from fluid accumulation in the tissues (edema). Warmth might indicate inflammation; coldness might mean restricted blood flow. The therapist attends to the pliability of the fascia—the thin fibrous sheets covering and traveling between muscles that extend way beyond the neck to regions as far as the hips and legs. The network of fascia all over the body acts as a flexible scaffolding that constantly mediates the transmission of muscular forces to the rest of the body. Normal movement and stretch in the fascia keep it in a soft gel-like state, but prolonged immobility causes it to congeal in a more taut, tensile state. So therapists press on the fascia to understand how tension is distributed within the neck and how it propagates from other body regions.

The pressures applied by therapists to the neck relieve discomfort and enhance range of movement at the very superficial touch of the skin. Neurons that transmit the sense of touch on the skin connect with pain-related neural circuits in the spinal cord and inhibit the transmission of pain signals from spinal neurons to the brain. The therapist's deeper pressure on the body causes muscle fibers to stretch and relax. Pressure can also change the physical properties of the underlying fascia, including their viscosity and elasticity, by causing detachment of molecular cross-links within the fibrous layers.

The effect of all this touch can be transformative. In his practice Raymond has noticed that after neck massage, clients commonly show signs of relief throughout their body, not just in their neck muscles. "The neck is connected to everything," he notes. "Often, when clients get off the massage table, they are just so different than when they walked in. Their breathing changes, their facial expression changes, their voice changes, their standing posture changes. Their balance is better, and their headaches are gone."[5] The world just seems a better place. In this sea of twenty-first-century worries, the primitive impulse to protect the neck is so strong and deep-seated that we subconsciously channel our contemporary distress—from difficult coworkers to frustrating urban traffic—into muscular tensions at the neck. But as one consolation, the hands of another person can dissipate this tension in our necks, and the sense of release can radiate back out all over our bodies.

• • •

We have all felt the hair on the back of our necks stand up. This reaction is likely an evolutionary holdover from our much furrier mammalian ancestors. When many mammals are threatened or backed into a corner, they raise their hackles. By erecting the hair around their head, they increase their apparent size as part of the "fight" option in the fight-or-flight response. This instinctive response is activated by the sympathetic nervous system that sparks the multipronged involuntary response to acute threat.

At a slightly more conscious level, humans have an additional response at the neck to alarm. We commonly touch our necks with our hands when we are endangered or just really nervous. When presented with shocking news or accused of wrongdoing, women often raise their hand to cover the suprasternal notch—the "neck dimple" at the juncture of the base of the neck and the collarbone. In these circumstances, men tend to actively rub the back or side of their neck or lightly grip their throat. Most any stimulus that elicits a fight-or-flight stress response can prompt neck-touching behaviors. Author Joe Navarro first noticed these behaviors and their gender differences in his college anatomy class when students initially confronted the disturbing sight of their dissection animals. Later, after he began his career as a counterintelligence agent, he used such observations to identify people under psychological distress.[6]

Navarro worked for twenty-five years at the FBI, where he devoted a lot of attention to decoding the nonverbal signals of informants and suspects. He found that one of the most reliable indicators of duress was self-touching the neck. To assess someone's affective state, writes Navarro, "keep your eyes on the hands and, as feelings of discomfort and distress surface in people, their hands will rise to the occasion and cover or touch their neck." This behavior once helped him locate a dangerous fugitive. In this investigation Navarro and his partner visited the criminal's mother at her home. When the mother answered the door, she was clearly tense, but she answered their questions about her son without hesitation or apparent alarm. At one point Navarro bluntly asked, "Is your son in the house?" She answered "no" while at the same time raising her hand to cover her suprasternal notch. With suspicions raised, Navarro continued the interview. Twice, he asked if the son might be in the house without her knowledge, and twice she denied the possibility while reaching to cover her neck. Struck by the consistency of this gesture and denial, Navarro asked permission to search the house. They found the young man hiding in a closet under blankets. A gesture at the neck was the giveaway.

Psychologists have systematically documented such neck-touching behavior in a range of social situations. In interactions between people of differing status, the more vulnerable person tends to touch their neck more frequently. For example, researchers at Harvard University found that during simulated job interviews, the applicants touched their necks and upper torsos significantly more than the interviewers.[7] Covering the neck in response to fear may seem like a defensive action, but Navarro argues that this neck-touching is more about self-soothing than self-protection. It is a subconscious pacifying behavior that may quell the defensive actions of the sympathetic nervous system and activate the calming actions of the parasympathetic nervous system by stimulating the densely innervated skin of the neck. In essence, it is a form of light massage.

SURFACE DEFENSES: INSULATION, SHIELDS, AND REMEDIES

Our evolutionary past has endowed humans with innate immune responses and subconscious neural reactions that have been crucial to our survival but that are largely beyond our control and intention. Evolution has also given us a brain capable of inventing devices to protect our necks and agile hands to materialize these inventions. For centuries, artisans and factories have crafted objects from textiles, steel, and plastic worn to buffer us from specific threats to the neck. But the designers of these guards forever face the conflict between protecting the neck and keeping it mobile and comfortable.

People living with cold winters commonly confront a seasonal threat to the neck, a body region that is particularly thin, lean, and vascularized, and thus a prime spot for heat loss. We can reduce this heat loss to a small degree with high-collared shirts like turtlenecks, but their thin loose collars only marginally insulate the neck. Instead, when we venture out into the frigid weather, we commonly wear thicker, bulkier scarves that can be removed easily when we reenter our heated buildings. Scarves vary splendidly, but there is one universal design constraint: they must be soft. Our tender necks cannot tolerate anything else. So we insulate our necks with scarves knit or woven from the softest wool, cotton, and silk.

• • •

Unlike some other body regions with vital organs—for example, our head and our thorax—the neck has no bony armor to protect it. Such a rigid skeletal shield at the neck would greatly limit head mobility, which has evidently been much more crucial for survival than even an armored neck. So, for nearly all daily activities, we leave our necks unshielded, relying on common sense and good luck to avoid harm. However, for our most dangerous activity—warfare—humans throughout history have devised a multitude of neck guards that afford at least some degree of head movement. In the early Middle Ages, beginning in eleventh-century France and Germany, European warriors commonly wore hoods composed of small, interlocking metal rings, termed "mail coif," that draped down to guard the neck while still permitting its movement. Such metal mesh limited the damage from slashing swords and spears. With the invention of more powerful projectile weapons such as longbows and crossbows, soldiers needed more impenetrable armor, and they protected their necks with solid metal plates extending down from the helmet or up from the chest armor. These early rigid shields, termed "gorgets" (derived from the French word *gorge*, meaning "throat"), often restricted head movement. In later versions gorgets were formed from articulated segments of steel that permitted more flexibility. But even these proved ineffective in protecting against the invention of firearms (guns and cannons), so they were largely abandoned as armor. Still, gorgets were retained as parts of ceremonial military uniforms as sites of elaborate engravings and decoration.

Rigid collars reappeared in the combat uniforms of British and American soldiers in the eighteenth century when they were made from thick leather and fastened with buckles in the back, providing protection against slicing cutlasses. In contrast to previous trends in military neckwear, the primary purpose of this stiff leather collar was to intentionally restrict head movement, enforcing a soldier's proper posture with his head forward and erect. In the United States the practice of wearing such collars ended in the 1870s, but the term "leathernecks" has carried on as a moniker for the US Marines.

With the invention of extremely strong fabrics, such as Kevlar, in the late twentieth century, warriors are now issued flexible and nearly impenetrable collars that protect the neck against shrapnel, which is the greatest threat to the neck in contemporary warfare.[8] Despite these new materials, the incidence of neck wounds in recent US wars (in Iraq and Afghanistan) was high in part because compliance in wearing the neck armor was low.[9] Soldiers commonly cited discomfort and interference with specific activities (e.g., aiming a gun from a prone position) as reasons for forgoing the collars.[10] For all the advanced engineering devoted to military technology, designers still have not resolved the trade-off between protection and mobility at the neck.

Certain injury-prone sports also use gear for neck protection. In baseball, catchers and home-plate umpires commonly dangle throat guards from their facemasks to block wayward pitches and foul balls. Many youth hockey leagues now require their players to wear padded collars that strap around the neck to protect against fast-flying pucks and dangerously sharp skates. While such equipment protects the throat from direct impacts and cuts, we have no such devices to protect the cervical spine and spinal cord against impacts, which are the source of the most serious trauma in certain sports (e.g., diving, rugby, horseback riding, skiing, cycling, and American football).[11] Stabilizing the neck to prevent such whiplash-like spinal injuries would necessarily and prohibitively restrict head mobility. For thrill, fun, or competition, we risk the neck.

Humans have devised all sorts of topical treatments to protect and soothe the neck. As a region that is typically exposed in warm weather, we slather the neck with sunscreen to guard against UV damage from the sun and spray it with insect repellent to ward off irritating or disease-carrying bugs. When our neck aches, we may apply creams that elicit a soothing sensation of warmth or cold, and when our throat is sore, we might gargle with a solution containing a mild topical anesthetic. In the past, humans formulated an even broader array of ointments and salves to treat the neck and throat. In the first century CE, the Roman naturalist Pliny the Elder devoted three chapters to neck remedies in his *Natural History*, which at the time was the most comprehensive pharmacopeia and encyclopedia of nature in the Western world.[12] For example, in one chapter Pliny offers a treatment for mucus accumulation and soreness in the throat. "Millipedes, bruised with pigeons' dung, are taken as a gargle, with raisin wine; and they are applied, externally, with dried figs and nitre [a salt found in ashes]."[13] In another chapter he recommends "the urine of a she-goat, injected into the ears, ... [or] a liniment made of the dung of that animal, mixed with bulbs," for cramping and stiffness of the neck.[14] We will probably never know the effectiveness of these treatments, but we can never question their inventiveness.

SUPERNATURAL DEFENSES: SAINTS AND MYTHS

Despite all our protective instincts and inventions, we humans remain deeply aware of our vulnerability at the neck. Try as we might, our fate is never fully in our control. Throughout history people have turned to supernatural powers to protect and heal their necks. Sometimes these are quiet, private rituals; sometimes they are events of the masses. Every February 3, for instance, the streets of Dubrovnik, Croatia, fill with thousands of people celebrating their city's patron saint and protector, Saint Blaise, a fourth-century Armenian physician-turned-bishop.[15] The tradition dates to the tenth century when, according to legend, a vision of Saint Blaise appeared to a local priest to warn the city of an impending attack by the Venetians, and Dubrovnik successfully resisted conquest. One highlight of the festival, now a UNESCO

Intangible Cultural Heritage event, is an annual procession containing the relics of Saint Blaise encased in precious metals and decorated with jewels. During the procession some in the crowd reach out to touch or kiss the relics as they are carried through the streets. Inside the ornate encasements are Saint Blaise's head, hand, foot, and notably his throat.[16]

Even before Dubrovnik adopted him as their own, Saint Blaise had been the patron saint of the throat for Catholics. On this same day in Catholic churches around the world, millions of parishioners line up to get their throats blessed by their priests. The supplicants kneel, and the priest touches them one by one with two crossed candles, reciting the prayer: "Through the intercession of Saint Blaise, bishop and martyr, may God deliver you from every disease of the throat." When Romans began persecuting Christians in Armenia, Blaise retreated to a cave but was eventually captured and taken before the governor of Cappadocia (in modern central Turkey). On the way a desperate mother rushed to Blaise and pleaded with him to help her young son who was choking on a fishbone. Blaise placed his hands on the boy's throat, prayed, and the boy was healed. Ironically, Blaise was ultimately martyred in 316 CE by beheading—a fatal severing at the throat and neck.

As noted earlier in the chapter, our biological protection against many infections comes from our immune responses, and the lymph nodes in the neck play a crucial role in this defense. However, the lymph nodes themselves are subject to disease. One such disease, scrofula, drove people for centuries to seek divine healing, delivered through the miraculous touch of their king or queen.[17] Scrofula is an infection of the lymph nodes by a tuberculosis-causing bacterium that manifests as bulbous masses or abscesses that protrude from sides of the neck, sometimes causing considerable disfigurement. In its most serious form the bacterial infection migrates from the cervical lymph nodes to the lungs to cause lethal pulmonary tuberculosis. Scrofula became prevalent in the medieval period as people began living in close quarters with their domestic animals, particularly cows, which transmitted the disease to humans. Some medieval physicians believed that scrofula was caused by sins such as gluttony, and they prescribed dietary restrictions.

A more popular treatment for scrofula was to seek the Royal Touch. Beginning with the Anglo-Saxon king Edward the Confessor (1042–1066 CE), monarchs of both England and France claimed to cure diseases with their divinely inspired touch. Scrofula was the most common affliction brought before the royalty. The Royal Touch became a mass healing ritual, reaching its pinnacle in the seventeenth century. In a large public ceremony on Easter Sunday in 1608, King Henry IV of France placed his curative hand on 1,250 subjects suffering from scrofula; and King Charles II of England, over his twenty-six-year reign (1660–1686), touched more than ninety-two thousand people, with the afflicted coming from as far away as Russia and the New World. During the Royal Touch ceremonies, patients knelt one-by-one before the king or queen, who stroked the patients' necks while the chaplain read the scripture "They shall lay their hands on the sick and they shall recover" (Mark 16:14). Each supplicant also received a gold coin, stamped with the image of Archangel Michael slaying a dragon—a symbol of victory over the evil malady. The coin was threaded with a ribbon and worn around the neck to ward off further afflictions.

People with scrofula commonly underwent spontaneous remission, and their reports of miracle healings following the Royal Touch were surely welcomed, if not encouraged, by royalty hoping to give credibility to their claim of divine powers and to legitimize their absolute monarchy. News of even a few instances of apparently miraculous cures was good public relations, and the pageantry of the Royal Touch was performed by kings and queens for nearly seven centuries. By the eighteenth century the idea of Divine Right monarchy waned, and the royal healing ceremonies became less popular and frequent. In the final years of the Royal Touch in England a desperate mother traveled three days by coach to London with her two-year-old son so that he could receive the Royal Touch from Queen Anne. In a ceremony that was among the last ever held, the boy received the Royal Touch along with two

Figure 20. Queen Mary I "healing" scrofula by touching the neck, ca. 1550. Copy of a sixteenth-century illustration from *Queen Mary's manual for blessing cramp rings and touching for Evil.* Painting by H. Hayman, 1916, the Wellcome Collection (Wikimedia).

hundred others afflicted with scrofula. Her son, Samuel Johnson (1709–1784), grew up to become one of England's most heralded poets, essayists, and critics. Despite the Royal Touch, however, Johnson never recovered from the disease, and it eventually left him blind in one eye and deaf in one ear. Nevertheless, until the day he died, Johnson wore around his neck the Angel coin, a touch-piece of his childhood affliction.[18]

In turning to Saint Blaise or the Royal Touch, people have sought divine intercession to heal their frail necks. But in other traditions people have created myths in which deities and mythological creatures have necks that are especially strong, and they use them as agents of protection. In one Hindu myth the god Shiva, who also goes by Nilakanta, meaning "blue-necked one," uses his throat in a great act of protection. In this myth the whole world was once threatened by a wave of deadly poison vomited by Vasuki, a snake king, or rising out of the churning Milky Ocean. Witnessing this imminent catastrophe, several gods petitioned Shiva for help. Shiva quickly drank up all the poison to save the world. Shiva's wife, the goddess Parvati, rushed to Shiva and clutched him by the throat to prevent the poison from entering his stomach and the rest of his body. The poison absorbed into Shiva's throat and turned the skin blue. Artistic depictions of Shiva commonly show his throat colored blue in tribute to his beneficent and protective act at the neck.

REPTILIAN DEFENSES: DECEPTION, COUNTERATTACK, AND CONCEALMENT

Just as in humans, the neck of other animals is a locus of great vulnerability, and some species have evolved structures and behaviors at the neck for actively defending against predators. Some of the most fantastic of these antipredator defenses are found in reptiles, especially lizards, snakes, and turtles. Many lizard and snake species use their necks to deceive their predators or counterattack with physical or chemical defenses. In the simplest but perhaps most effective form of defense, turtles just tuck their head away within their body armor. The particular profusion of these antipredator adaptations among reptiles may be related to the distinct combination of anatomy, behavior, and physiology underlying the reptilian lifestyle. Reptiles are quite diverse, but most are small and move around during the day as lone individuals, making them especially vulnerable to large visual predators. At the same time, their near-to-the-ground posture and their limited aerobic capacity greatly limit reptiles' ability to make long-range escapes. They can rarely outrun their attacker. On the fight-or-flight continuum, reptiles must more commonly resort to fighting.

Perhaps the most iconic antipredator display at the neck is found in the Australian frill-necked lizard.[19] These slender lizards generally inhabit arid environments, and their brown and gray coloration blends in well with their surroundings. When approached by a predator, usually a bird or snake, they turn to face their enemy, open their mouths widely, and rapidly erect their large colorful neck frill—an expandable sheet of folded skin supported by greatly elongated bones of the throat (hyoid) skeleton. Both the interior of the mouth and the expanded frill are brightly conspicuous, with patches of yellow, red, orange, and white. In this defensive posture the lizards also commonly hiss, sway their bodies, and whip their tails. The onset of the display is rapid, less than half a second, and the open frill is impressively large (about 25 centimeters [10 inches] in diameter). Opening the frill increases the apparent size of the animal's head more than fivefold. The startle effect on the predator alone may be enough to deter or delay a strike. But for predators that attempt to strike, the frill can also confuse the predator about where exactly to aim their attack to capture the lizard by the head. A bite to the frill is likely just an injury, while a bite to the head would likely be lethal.

Another familiar case of a reptile using its neck to exaggerate its size is the defensive posture of cobras. When cobras are threatened, they often vertically elevate the head end of their body and laterally expand

Figure 21. Frilled lizard showing the expansion of its neck frill in a defensive posture. Photograph by Warren Garst, 1972.

the neck region to form a broad hood. In some cobra species the expansion of the neck also reveals warning patterns. To erect the hood, one set of muscles spanning from the neck vertebrae to the ribs causes the ribs to swing forward to broaden and flatten the neck region while other muscles attaching the ribs to the overlying skin contract to keep the skin taut.[20] If the threat persists, some cobra species may spit out a potent defensive toxin from their fangs or lunge toward the predator.

Rather than expand their necks, coral snakes commonly sharply bend their necks to deter a lethal attack. When threatened, they kink their neck in either direction while thrashing their tail erratically. The tail motion presumably draws the attention of the predator away from the head while the kink confuses the predator about the position of the head. The thorny devil, an Australian lizard, takes predator deception even further. When confronted by a predator, the lizard tucks its true head between its front legs and presents a spiny bulbous growth on the back of its neck that is sometimes termed a "false head." This posture likely deceives the predator into attacking a fake head, and the large thorns facing forward in this defensive position make the lizard especially difficult to grab and swallow.[21]

Wrynecks are birds, not reptiles, but in a remarkable display of mimicry they ward off their predators by imitating deadly reptiles. They have brown and black splotches and streaks on their back that, when viewed from above, resemble the color pattern of vipers living in the same habitat. Their name—the wryneck—derives from their peculiar neck motions that are elicited by a predator. When disturbed, these birds hiss while torquing and twisting their neck in all directions in a writhing motion that eerily mimics the movements of a viper.

• • •

Turtles lumber slowly through the world with the seeming confidence that they have nearly impenetrable defenses. They live in fortresses. Their shell and the protection it provides against predators are the cornerstone features of a turtle's stable lifestyle. Indeed, in certain Native American and Hindu traditions the whole earth rests on the back of a turtle. Their rigid shell, composed of vertebrae and ribs fused with dermal bone, makes for solid armor, but it greatly restricts any movement of the thorax and backbone. The shell is also heavy, accounting for over one-third of the body weight in many turtles, further limiting the possibility of agile locomotion.

As partial compensation for the restricted mobility in their body, turtles' necks are highly flexible. This flexibility allows them to move their head in a wide arc to feed and collect a broad range of sensory information even while they are stationary or moving at a turtle's pace. Moreover, their flexible neck allows them to do what no other vertebrate animals can do. They can completely hide their head and neck within their body. Unlike the highly flexible necks of other vertebrates (e.g., long-necked birds), the necks of turtles are not especially well endowed with a large number of mobile parts. All turtles have eight cervical vertebrae, far fewer than long-necked birds (twelve to twenty-five cervical

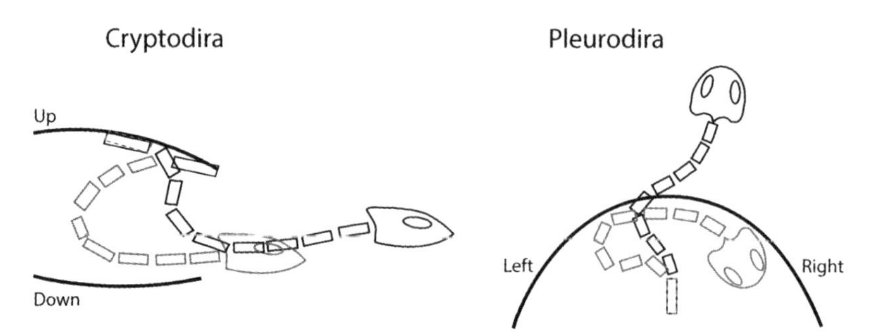

Figure 22. Neck retraction mechanisms in the two major groups of turtles: side-view of a cryptodiran (hidden-neck) turtle and top-view of a pleurodiran (side-necked) turtle. Illustration in Anthony Herrel, Johan Van Damme, and Peter Aerts, "Cervical Anatomy and Function in Turtles," in *Biology of Turtles* (Boca Raton, FL: CRC Press, 2007), 177–200.

vertebrae) and barely more than the seven found in mammals, whose necks could never bend so much. Moreover, the shapes of turtle cervical vertebrae are not so unusual. Instead, much of the flexibility comes from the fact that their neck vertebrae are connected with rather loose joints that can routinely dislocate to permit extreme bending.

Within their ancient body plan, turtles retract their necks in one of two ways. Indeed, the mechanism of neck retraction is the primary distinguishing feature of turtle evolution, dividing turtles (order Testudines) into two suborders: Cryptodira ("hidden neck") and Pleurodira ("side neck"). Hidden-neck turtles, which include tortoises and the familiar pond turtles of North America and Europe, retract their head by bending the neck primarily at two cervical joints, one joint in the middle of the neck and one at the base, to make a vertical S-fold in their neck. Side-necked turtles, which live only in the Southern Hemisphere, retract their head by bending the neck laterally mostly at a single cervical joint near the base of the neck and swinging the head sideways under the shell.

In the evolutionary past the earliest turtle species probably had little ability to retract their necks. The simpler side-necked mechanism of Pleurodira evolved first, followed by the independent evolution of the double-bend mechanism of the hidden-necked Cryptodira. Looking at living turtles, it seems obvious that the function of these retraction mechanisms is to protect the head and neck. Such protection was indeed probably an important selective advantage driving the early transition from no retraction to side-neck retraction. But the independently evolved double-bend retraction mechanism likely originated for a different purpose. Fossil evidence indicates that pulling the head in a vertical S-bend was originally less about hiding the head away from predators and more for positioning the head for prey capture.

One of the earliest documented species of turtles, a 150-million-year-old fossil, *Platychelys oberndorferi*, was in the same lineage as side-necked turtles, but careful biomechanical analysis of its vertebrae indicate that it probably folded its neck vertically.[22] However, this bending was not enough to hide the head, and evidently this early experiment in vertical neck retraction was for something other than protection. Other features of the cervical vertebrae in this fossil turtle were similar to those of two modern turtles, the alligator snapping turtle and the matamata, who use their necks for more than hiding their heads. These two living turtle species use their necks to feed with rapid explosive predator movements; they are ambush and suction feeders. They spend much of their day on the bottom of the lake disguised and waiting, with their necks retracted. When a fish swims by, they quickly thrust open their mouths and enormously expand their throats while lunging their head forward to suck in their prey. If the earliest hidden-neck turtles fed with such action, as their fossil neck anatomy suggests, this means that the original function of its S-bent neck was not to pull its head in for protection but to cock its neck rearward preceding a rapid forward thrust. Once this feeding mode evolved, neck morphology secondarily evolved to enable full hidden-neck retraction and thus protection of the head. So, for most living turtles, their neck that serves them so well in their slow, defensive lifestyle may have in fact arisen for high-speed predation.

• • •

It's hard not to envy turtles. Their calm lives seem the antithesis of the hectic lives many humans live. They commonly bask in the sun while we scurry around full of worry and exhaustion. We cannot be certain about the internal worlds of turtles, but just looking at them, it is difficult to imagine that they end their days with sore necks, thinking up inventions to insulate and shield their necks, or anxiously appealing to higher powers for healing. Their calmness must come, at least in part, from the security of carrying their own refuge on their back. And yet their shells are only fully protective because they have thin and extremely flexible necks that enable them to hide their heads. I can't help but wonder if turtles sense that their neck is their protector more than their vulnerability.

For humans, with our thin necks that we cannot hide, we are fated to carry a sense of frailty at the neck. There, we are susceptible to injury, pathogens, and cold, but we cannot carry the burden of protecting ourselves in our consciousness all the time. Instead, we rely heavily on physiological protections beneath or at the edge of our awareness or on supernatural protections beyond our control. Somewhere in between, we act as we can, turning to warm tea, soft scarves, and soothing massage for comfort and solace.

EPILOGUE

Created & Crafted

The Necks of Pottery Vases

The many and diverse necks explored in this book are fundamentally organic structures. They are products of the wildly creative forces of natural and sexual selection acting over eons and the quirky generative forces acting in the embryo over days and months. Their construction always proceeds through modifications of ancestral blueprints, and they are built from a restricted set of materials—bone, sinew, muscle, and epithelium—that have intrinsic physical limitations. Their design is further constrained by the necessities of performing many distinct, sometimes competing functions. Any neck must be a jack-of-all-trades and thereby a master of none. Necks, like all organic structures, are shaped by elaborate histories and bounded by internal limitations. But not all necks arise organically. Some are conceived in our imaginations and created with our hands. Unbound from the constraints of organic processes, humans expand the possibilities of neck-like forms and incorporate them into crafted objects, such as ceramic vessels. Here, I close the book with considerations that opened the book and that started me thinking broadly about necks in the first place. What shapes the necks of pottery and what do these inanimate necks reveal about the design and meaning of living necks?

Compared to the necks of animals, the necks of ceramic vessels are formed quickly. Working at the potter's wheel, the potter applies an inward and upward force onto wet, hollow, spinning clay, and the whole genesis commonly occurs within minutes. In contrast to biological material, clay is a homogeneous material, so in some ways the physics of clay is comparatively simple. However, when building with clay, the potter is always on the clock because the mechanical properties of clay change drastically as it becomes first wetter and softer and then drier and more brittle. Through skilled touch and careful timing, potters have stretched the necks of vessels into an impressive variety of shapes. Still, ceramic necks, like those of animals, have structural limits.

In addition to structural constraints, many vases have had to meet certain functional demands. (I use the word "vase" as a shorthand for any ceramic vessel with a constriction that leads up to the rim.) In contrast to animal necks, which epitomize multifunctionality, the necks of vases perform only a small number of roles. For nearly twenty millennia, vases served two primary functions—storing and dispensing—both of which rely on the size and shape of the neck.[1] As storage vessels, vases have narrow, sealable openings that restrict the entry of the unwanted—for example, air and heat that could degrade stored wine or olive oil and pests that might eat stored grain. Throughout history, most vases have stored consumables, but some have also been repositories to protect precious items, such as the ashes of the dead or sacred texts. The design of many vases reflects their role in dispensing liquids. Vases that poured a large volume, like the wine-storing amphora and water-storing hydria of ancient Greece, tended to have shorter, broader necks, while those that dispensed precious perfume or funerary oils, such as Greek lekythos, had proportionally longer, narrower necks that could pour their liquids with a more precise, controlled flow. In more recent centuries vases have commonly been used to display flowers. Bud vases that hold only a few stems usually have long and narrow necks, while vases supporting wide floral sprays are often stouter for stability, with broader necks to hold many stems. In all these vases, form must follow function.

While vases have retained some utility in modern times, they now serve mostly an aesthetic purpose. They just stand there and look beautiful. Their curvatures and proportions are apparently endlessly pleasing to human eyes. Formed in all sorts of sizes and shapes, vases have been among the most common and treasured pieces of sculptural art throughout history and across the globe. Within the diversity of shapes, each portion of the vase can vary in width, height, and arc, but to please the viewer, the parts cannot vary independently. They must combine to flow together harmoniously. In this regard, vases embody the aesthetic principle of "unity in variety," the idea that people perceive objects as beautiful when they have "different parts (variety) that are related by some common feature or can otherwise be conceptualized as a coherent whole (unity)."[2] The necks of vases can be long and narrow as an oboe or short and wide as a tuba, flaring as a French horn or straight as a flute, but each neck variation must fit within the greater composition of the pot.

But what specifically about the form of a vase makes it so appealing? In 2023, German researchers presented a group of sixty participants a series of twenty-five mathematically-defined profiles of vases and asked the participants to rate their beauty.[3] Researchers found that neither curvature nor proportion alone could predict how the vases were rated, but taken together, these features could account for nearly all the variation in perceived beauty. At least within this population, the discriminating eye simultaneously weighs shape and line. Both contribute to the unity within the variety. For potter George Pearlman, proportion and curvature come together at transition points within a vase at junctions between the lip, the neck, the body, the foot, and the ground. "All the important decisions in making a vase happen at these transition spots," says Pearlman. "It's at those junctures that you start perceiving proportions."[4]

As a major transition zone in the vase, from the lip to the body, the neck more than any other region conveys the vessel's overall proportion in a single glance. According to Pearlman, the interface between the air and the upper rim, the lip, is the critical invitation to the viewer. It is the defining edge and "welcoming spot" for the eye. So a primary aesthetic function of the neck is to give prominence and support to the lip. If the lip is what catches the eye, it is the neck that starts the eye moving downward, drawing it across the shoulders and on to the body. For Pearlman this flow is the essence of grace as an aesthetic value. As he explains: "There is also something specifically about the vertical dimension of this flow that is a component of grace as well. Necks of humans are graceful because they require active balance to keep the posture upright, and the necks of vases are graceful because, in their static stance, they allude to this active balance." The vase's neck is also a passage to the interior, to the invisible. "As a transition to that space, the neck makes that volume more mysterious." Pearlman likens vases to the earliest known structures of human habitation: caves. In these dwellings, the back of the cave was the most sacred space, but the area just inside the mouth of the cave—the neck—also had special significance. It was often decorated with precious objects or paintings that welcomed residents or warded off intruders. By analogy, the neck of a vase serves as a transition zone to the sacred space within the vessel.

Vases are more than shapes; they are constructions. Pearlman has been throwing pots for more than forty years, most of that time at his studio in coastal Maine. Since 2014 he has focused on tall, statuesque porcelain vases, some of which have been exhibited at major museums and galleries across the United States. Many potters over the millennia have constructed such large vases in separate pieces—for instance, the foot and body as one thrown piece and the neck and lip as another piece—and then joined them together after they partially dry. Working with smaller, firmer pieces in this way alleviates some of the technical challenges of forming a long, constricted neck. But Pearlman throws his large vases in a one-act drama, the entire pot created as a continuous gesture. Forming the neck is typically the most precarious, climactic phase of making a vase. In the same way that humans must be especially gentle in touching each other's necks, potters must also work with supple fingers when forming the neck of a vase. If the pressure is

Figure 23. Ceramic vase by George Pearlman, 2019. Photograph by George Pearlman.

too strong or uneven or if the clay is not adequately lubricated with water, the neck can easily torque or rip from the shoulders. But, for Pearlman, the precarious shaping of the neck is a moment of delight. "There is just something so thrilling about holding all that tension."

Pearlman recognizes that the necks of vases, like those of humans, are a locus of vulnerability. In a recent series of pots he systematically explored this vulnerability at the neck, and in doing so, highlighted a fundamental structural difference between humans and pots. Humans have an endoskeleton, with our heads and necks supported on an internal column, the spine. But pots have an exoskeleton. The weight of their upper end—their "heads," lips, and neck—rests on the shoulders with the force of gravity transmitted by the circumference of the pot's neck. There is no internal structure connecting it to the lower body, as in humans. Pearlman probed this different kind of structural vulnerability by deforming the vases' shoulders from the inside while they were still moist and pliable and observing what happened to their necks. If he stretched the shoulders too much, the neck would completely collapse. With less drastic deformations, the neck would gently slump but remain mostly upright. These vases, caught in the process of distortion, hinted at the neck's universal vulnerability but one that depends on a vase's distinct form of structural support.

Despite these structural differences, Pearlman and many other ceramic artists see vases as a metaphor for human bodies. Their parts even have anatomical names: lips, necks, shoulders, bodies and feet; some vases may also have waists and hips. For Kenyan-British ceramic artist Magdalene Odundo, staying connected to human anatomy is just a matter of routine daily observations. "Every day I go into the studio and when I start making a piece of work, … [I] look at how people are interacting with whatever they are doing.… I love thinking of that relaxation and the breathing that this person is emitting: breathing in and out, creating that vessel in the body."[5] In describing her process, Odundo gives particular significance to the neck and "throat" of the vessel. After forming the bulge at the base, "then you touch the neck and open the rim so that [the vessel] becomes more human and animated." The neck confers the vase's vitality.

For Korean potter Heo Jin-gyu, making the neck of his large kimchi pots (*onggi*) is the peak moment in their creation.[6] In some ways it is the culmination of processes that began in his earliest days—even before birth. As someone whose mission is to preserve the cultural legacy of kimchi and its pottery, he believes his destiny as an *onggi*-maker began in the womb, when his pregnant mother helped his father make *onggi*. In constructing the pot, he begins with a round slab pressed onto a foot-powered potter's wheel. He builds cylindrical walls from thick coils and then flares the walls outward as he thins them with a pair of hand-crafted wooden tools. Finally, he arches the top half inward to make the constriction, the *jeon* or neck. "When we are making the neck of the pot, it's like the final touch. When we go hiking, reaching the peak is the best feeling. That's how I feel when I am making the neck, like I've reached the top of the mountain. That's why I put a lot of work into making a beautiful neck."

• • •

Vases and the necks that define their form are indeed beautiful. They embody so much of what we find appealing in proportion and curvature. Their human scale draws us in and enables us to see ourselves in their form. This palpable humanness of their proportions and curvature was what prompted me to explicitly compare them to animate necks decades ago, and what has drawn me to make them and delight in them ever since. However, as resonant as vases are as metaphors for human shape, I have come to see that they are perhaps not such apt metaphors for life. Vases are static, hollow, and composed of a single material, and surely life, if nothing else, is essentially dynamic, dense with working parts, and built from an amalgam of materials—as in the neck illustration on the cover of the *Gray's Anatomy* my dad gave me years ago.

All life, including that short segment between the chin and collar, is filled with complex conglomerations and motions, glorious structures and behaviors refined through millions of natural experiments. Life's diversification and persistence over eons attest to an inexhaustible vigor across generations. At the same time, creatures die as surely as they live and procreate. For better or worse, we humans pass through this world with a deep awareness that our life-force is potent and regenerative but ultimately vulnerable and ephemeral. Much of what we do in our arts, religions, and in our own psyches are attempts to reconcile with this inevitable condition of existence. In this sense necks, with all their concentrated vitality and vulnerability, are a compelling metaphor for life.

ILLUSTRATIONS

NOTES

PREFACE

1. *Los Angeles Times* 1927.

2. Gray 1988.

CHAPTER ONE. WHY & WHAT

1. Daeschler, Shubin, and Jenkins 2006, 757.

2. Gans 1992, 17. For a more recent elaboration of this idea that incorporates information from key fossils found in recent decades, see Shubin, Daeschler, and Jenkins 2015, 63.

3. Some bony fish have circumvented this problem by a quick contraction of one side of the body into a "C-turn" or by protruding their jaws, sucking in prey rather than lunging at them.

4. MacIver and Finlay 2022, 2.

5. Plato 2009, 59.

6. Tauber n.d.

7. Aristotle 1882, book III, part 3.

8. Vesalius 1998, 57.

9. This division of the physiological systems and their neuroendocrine control is very blurry; almost all physiological processes regardless of their pace are influenced to some degree by both the nervous and endocrine systems.

CHAPTER TWO. POSTURE & POSE

1. Hansraj 2014, 278.
2. Fiebert et al. 2021, 1261.
3. Lieberman 2011, 59–61.
4. Lieberman 2011, 349.
5. Lieberman 2011, 348.
6. Leonardo da Vinci 1952, 111.
7. Lieberman 2011, 349.
8. Lieberman 2011, 361.
9. Lieberman 2011, 365–372.
10. "Hard labor, guy stacking 20 bricks on his head," www.youtube.com/watch?v=t1vDPcXTRIs&ab_channel=PieterTerpstra, accessed August 11, 2023.
11. Rockel 2006, 107–108.
12. Grant 1989, 63.
13. Grant 1989, 76.
14. Studies in the 1990s indicated that head porters in the Luo tribe in East Africa expend almost no additional energy when they are carrying heavy headloads compared to walking without loads (Heglund et al. 1995, 52). More recent studies have yielded more ambiguous results about the ergonomic advantages of head porterage.
15. Dave et al. 2021, 17.
16. Because the cervical spine actually has an S-shaped curve, vertical forces also impart some shearing forces that tend to push the intervertebral discs toward the back or the side, further threatening the necks of headloading porters.
17. Lloyd et al. 2010, 522.
18. Allen 2014, 1.
19. Haneline 2009, 119.
20. Sterling and Kenardy 2011, xii.
21. Sterling and Kenardy 2011, 16–24.
22. Sterling and Kenardy 2011, 9–12.
23. Carroll et al. 2009, 1063; Holm et al. 2008, 763.
24. Ferrari 2006, 7; Ferrari, Constantoyannis, and Papadakis 2001, 254.
25. Ferrari 2006, 4–8. For a review of this thesis, see also Haneline 2009.
26. Elliott, McMenamin, and Walton 2016, 7.
27. Chen et al. 2013, 1.

28. Ernstbrunner et al. 2017, 2125.
29. Chang et al. 2016, 12006.
30. Sharker et al. 2019, 1.
31. Aristotle 2011, book II, part III.
32. Bruno and Bertamini 2013, 1.
33. Schneider and Carbon 2017, 1.
34. Sedgewick, Flath, and Elias 2017, 3.
35. Nicholls et al. 1999, 1521.
36. Costa, Menzani, and Bitti 2001, 63.

CHAPTER THREE. PANORAMA & GESTURE

1. Not all neck muscles are simply for controlling head and neck movements. Some muscles extend below the neck to move or stabilize the shoulder, and a few extend above the neck to move the jaw, tongue, and facial skin.

2. Lieberman 2011, 338–344.

3. Paley 1854, 52.

4. "Chicken Head Tracking," www.youtube.com/watch?v=_dPlkFPowCc&ab_channel=SmarterEveryDay, accessed June 17, 2022; and "Chicken Powered Steadicam," www.youtube.com/watch?v=UytSNlHw8J8&ab_channel=SmarterEveryDay, accessed June 17, 2022.

5. Crawford 1964, 357–360.

6. Cole 1995.

7. Plato 2013, 107–110.

8. Some birds with little binocular overlap use active head movements to enhance depth perception. By wagging their heads back and forth, each eye captures a slightly different view of their world, and these different views over time enable them to estimate depth.

9. The 270-degree swivel of an owl's neck is shown in this video: "How Owls Swivel Their Heads," www.bbc.com/news/science-environment-21279609, accessed July 14, 2023. Some raptors accomplish omniscience in a different and rather comical way. They roll their head forward almost 180 degrees and look at the world behind them with a completely upside-down view.

10. Krings et al. 2017, 12.

11. de Kok-Mercado et al. 2013, 514.

12. VanBuren and Evans 2017, 608–626.

13. Walker 2011.

14. Dillon 2017.

15. Bitti 2016.

16. One notable exception to the general tendency in ballet to move the head subtly and slowly is the use of the neck in spotting while the dancer spins.

17. Conyn 1953, 41–43.

18. Anderson 1986, 158.

19. Ramya 2019, 31–37.

20. Interview with Rachna Ramya, April 14, 2022, Hartford, CT.

21. Ramya 2019, 12.

22. Aishwarya Chandrashekhar, reply to "What is the significance of neck movement in Indian classical dances?," Quora, www.quora.com/What-is-the-significance-of-neck-movement-in-Indian-classical-dances, accessed January 4, 2023.

23. Interview with Ramya.

CHAPTER FOUR. TUBES & TRANSPORT

1. Uematsu et al. 1983, 256.

2. These are conservative numbers based just on resting rates. If you figure in periods of physical exertion, the numbers would be even higher.

3. Rozsa 2021.

4. Carotid arteries also house a sensory structure, the carotid body, which detects the concentration of oxygen in the blood.

5. Scally et al. 2012, 172.

6. Morimoto et al. 2018, 1.

7. Seymour, Bosiocic, and Snelling 2016, 1.

8. Natterson-Horowitz et al. 2021, 249.

9. Liu et al. 2021, 1.

10. Kimani 1987, 257.

11. Sadhra et al. 2015, 1.

12. Blushing is not the only strong emotion broadcast from the vasculature of the neck. People in the throes of rage often show bulging neck veins.

13. Elflein n.d.

14. Aristotle 1882, 65.

15. Of course, we can also intake air through the mouth, and in this case the air still must be guided forward to move down the trachea.

16. Held 2009, 105 (emphasis in original).

17. Lieberman 2011, 59–63.

18. Some snakes solve the problem of simultaneously eating and breathing with a remarkable tracheal adaptation. When a snake eats, its prey often fills its mouth for a long period, which might block the flow of air into the trachea. However, in some snakes the trachea is protrusible and adjustable laterally. When their mouths are full, they move their trachea forward or beside the prey to the edge of their mouth so that the trachea can continue to transmit air to the lungs.

19. Goldbogen 2010, 127.

20. Gil, Vogl, and Shadwick 2022, 898–903.

21. Louchart and Viriot 2011, 663.

22. Contradicting a strict association between flight and a toothless skull, even flightless birds lack teeth. For a tall flightless bird, like an ostrich, with an elevated head that forages on the ground, it might be a relief to have just a light head to move up and down.

23. Grajal et al. 1989, 1236–1238.

24. Button et al. 2018, 12501.

25. Laguarta, Hueto, and Subirana 2020, 275–281.

26. Ang 2022.

27. Hale 2022.

28. Benjafield et al. 2019, 687–688.

29. Torre et al. 2019, 98–101.

30. Schneider 2020.

31. Cowell 2008.

32. Fountain 2012.

33. Fountain 2012.

34. *Nature* 2016.

35. Belluck 2021.

36. Taylor and Wedel 2013.

37. In the bird respiratory system half of the inhaled air goes to the lungs and half goes to one of the air sacs. The entry of the new air into the lungs pushes the "used" air from the lungs into another set of air sacs. At exhalation this depleted air in these second air sacs exits through the trachea while new air that had been held in the first air sacs goes into the lungs.

CHAPTER FIVE. PACE & SCAFFOLDING

1. Hovet 2022.

2. Smith 2022.

3. Collis 2022.

4. WHO 2005.

5. Zimmermann and Andersson 2021, R13–17.

6. Crockford 2009, 155.

7. Our phylum, Chordata, includes vertebrates such as fish, amphibians, reptiles, birds, and mammals, as well as protochordates, small marine animals that share most vertebrate features but have no backbone.

8. Okabe and Graham 2004, 17716–17719.

9. Parathyroid hormone and calcitonin also act on the kidney to regulate the retention or elimination of calcium through the urine.

10. Leung, Braverman, and Pearce 2012, 1742.

11. Sterpetti, De Toma, and De Cesare 2015, 591–596.

12. Thomas Wharton, who is credited with naming the thyroid gland in 1656, wrote that one function of the thyroid gland was to enhance feminine beauty. "[The thyroids] contribute greatly to the roundness and beauty of the neck," he wrote, "because they fill the empty spaces around the larynx, and reduce the protuberance of these parts generally to smoothness and regularity, especially in women, in which they occur larger for this reason, and render more uniformity and charm to their necks" (Lydiatt and Bucher 2011, 9).

13. Twain 1880, chapter 48.

14. Lee and Chiu 2021, 577–579.

15. Zimmermann 2008, 2061–2062.

16. Leung 2012, 1740.

17. Zimmermann and Andersson 2021, R13–19.

18. Zimmermann, Jooste, and Pandav 2008, 1251.

19. Kristoff 2008.

20. American Thyroid Association, "General Information" www.thyroid.org/media-main/press-room/#:~:text=An%20estimated%2020%20million%20Americans,thyroid%20disorder%20during%20her%20lifetime, accessed May 4, 2022.

21. The original work establishing the neural crest as the embryonic source of C-cells was based on experiments in avian embryos. Recent studies on mice have questioned this neural crest origin. Nonetheless, these mice studies showed that calcitonin-producing cells have an origin distinct from those producing thyroid or parathyroid hormone, and they become incorporated within the thyroid after a secondary migration (Johansson et al. 2015, 3519).

22. "George Crile Sr.'s 25,000th Goiter Operation Photograph," 1936, Ohio Memory Collection, State Library of Ohio, https://ohiomemory.org/digital/collection/p267401coll36/id/8618, accessed July 23, 2022.

23. Hannan 2006, 187–191.

24. Little and Seebacher 2014, 1642. Thyroid hormones also act to boost metabolic rate and locomotor physiology in ectothermic "cold-blooded" vertebrates, particularly when they are exposed chronically to cold temperatures. These actions might have been the evolutionary precursor to the ability of endotherms to maintain high body temperatures even when the environmental temperature is low.

25. Though less dramatic than frog metamorphosis, many groups of diverse animals that undergo major morphological transitions use thyroid hormones to coordinate the change. These include flatfish (e.g., flounder and halibut), sea urchins, sand dollars, oysters, and scallops.

26. Mughal, Fini, and Demeneix 2018, R160–186.

27. Johns Hopkins Medicine 2020.

28. Not surprisingly, perchlorate also greatly affects amphibians. At concentrations found in polluted sites, it significantly decreases metamorphic rate (Couderq, Leemans, and Fini 2020, 110779).

29. Broder 2011.

30. Friedman 2022.

CHAPTER SIX. WORD & FLESH

1. Clayton and Philo 2010, 50.

2. Clayton and Philo 2010, 51.

3. Lieberman 2011, 328.

4. Gross 1998, 216–217.

5. Because the two nerves loop around different arteries, their path lengths are somewhat different: about 130 centimeters (51 inches) on the left and about 60 centimeters (24 inches) on the right.

6. Wedel 2011, 251.

7. Gross 1998, 218.

8. Boë et al. 2019, 1.

9. Sasaki 2006.

10. Chen and Wiens 2020, 2.

11. Pentreath 2021.

12. Nooshin 1998, 70.
13. Kingsley et al. 2018, 10209–10212.
14. Riede et al. 2019, 1.
15. Elemans et al. 2008, 1.
16. Riede et al. 2008, 635.
17. Jakobsen et al. 2021, 2.
18. De Boer 2012, 1.
19. Nishimura et al. 2022, 760.
20. Boë et al. 2019.
21. Fitch, De Boer, Mathur, and Ghazanfar 2016.
22. Jarvis 2019, 50–54.
23. Joanne Scattergood, interview with the author, Simsbury, CT, July 5, 2022.
24. Colapinto 2022, 246–251.
25. Seashore and Metfessel 1925, 538–542.
26. Howes et al. 2004.
27. Seashore 1931, 623–626.
28. "Singer" n.d.
29. Sissom, Rice, and Peters 1991, 67–78. Recently, researchers have also shown that a cat larynx can produced purrlike sounds even in the absence of neural input and muscle contraction (Herbst et al. 2023, 4727).
30. Herbst et al. 2012, 595–599.
31. Pisanski et al. 2014, 89.
32. Pisanski and Reby 2021, 1–9.
33. Fitch 1999, 31–48.
34. "How" n.d.
35. Titze and Palaparthi 2018, 2813.
36. Podos and Cohn-Haft 2019, R1068–1069.
37. Williams 1986, 6–7.
38. Titze 2012, 52.
39. "Throat Singing" n.d.
40. Bergevin et al. 2020, 1.
41. Suthers, Goller, and Hartley 1994, 922–993.
42. Suthers, Vallet, and Kreutzer 2012, 2950–2959.

CHAPTER SEVEN. COURTSHIP & ATTRACTION

1. See the NIH guidelines here: Office of Research on Women's Health, "Sex and Gender," https://orwh.od.nih.gov/sex-gender, accessed July 2, 2023.

2. Darwin 1871, 521.
3. West 2005, 230–232.
4. West and Packer 2002, 1339–1343.
5. West 2005, 232.
6. West and Packer 2002, 1339. Other researchers found no difference in core body temperature associated with mane color or length but found that males with dark manes visited watering holes more frequently. Dark-maned males might restrict their activity to nearby watering holes to keep cool.
7. West and Packer 2002, 1340.
8. Wilkinson and Ruxton 2012, 619.
9. Darwin 1871, 502.
10. Simmons and Scheepers 1996, 771.
11. Simmons and Altwegg 2010, 7.
12. Wang et al. 2022, 1.
13. Sinervo and Lively 1996, 240–243.
14. Zamudio and Sinervo 2000, 14427.
15. Starnberger, Preininger, and Hödl 2014, 281–282.
16. Dudley and Rand 1991, 160.
17. Taylor et al. 2008, 1089–1090.
18. Taylor et al. 2011, 819–820.
19. James et al. 2022.
20. Zamponi et al. 2021, 692–693.
21. Markova et al. 2016, 88–89.
22. T'sjoen et al. 2011, 635–638; Cler et al. 2020, 748.
23. Puts et al. 2016.
24. Aung and Puts 2020, 154–155.
25. Reviewed in Aung and Puts 2020, 154.
26. Puts 2005, 388.
27. Puts et al. 2016, 2.
28. Reviewed in Puts and Aung 2019, 189.
29. Feinberg, Jones, and Armstrong 2018, 901–903.
30. Puts and Aung 2019, 189.
31. Feinberg, Jones, and Armstrong 2019, 192.
32. Pavela Banai 2017.
33. Pipitone and Gallup 2008, 268.
34. Fraccaro et al. 2011, 57. There is also evidence that women commonly lower the pitch of their voice in certain sexual contexts—for example, when

asked to produce a "sexy voice" or when addressing men whom other women have rated as attractive. Reviewed in Hughes and Puts 2021, 4.

35. Pisanski et al. 2018, 1.

36. Klofstad, Nowicki, and Anderson 2016, 284–288.

37. Borkowska and Pawlowski 2011, 55–56.

38. *NPR* 2011.

39. Holmes's deep-pitched voice can be heard in this TEDMed Talk on YouTube: www.youtube.com/watch?v=ywH-nbcCZfw, accessed June 14, 2023.

40. Chozick 2023.

41. According to Harold Koda, curator at the Metropolitan Museum of Art and author of *Extreme Beauty: The Body Transformed*, "the preference for a long neck is perhaps the only corporeal aesthetic that is universally shared . . . in all cultures, the head held high is associated with dignity, authority and well-being" (quoted in Mirante 2006). However, in my informal survey of world art, this apparent preference for long necks in portraits of women has notable exceptions. For example, women in classical Indian art commonly have short, broad necks.

42. Zheng et al. 2013, 899–903.

43. Moore 1985, 237. Separate but similar studies reported that courting men do not show these movements. Renninger, Wade, and Grammer 2004, 416.

44. Wickler and Seibt 1995, 402–404.

45. Preston-Whyte and Morris 1994, 20–40.

46. Achebe 2018, 123.

47. Achebe 2018, 119–142.

48. Schoeman 1983, 151.

49. "Northern Nguni Zulu Beadwork—Courtship Meanings," Museums Victoria, https://collections.museumsvictoria.com.au/articles/16600, accessed June 27, 2023.

50. MacDonell 2022.

51. Pointon as quoted in Biondi 2022.

52. Damas 2018.

53. Williquette 2019, 37–38.

54. The particular eroticism of the neck is borne out in quantitative research on erogenous zones. In a survey of almost eight hundred men and women of diverse races, ages, nationalities, and sexual orientations, the neck was ranked the third most erogenous area, trailing only the genitals and mouth (Turnbull et al. 2014, 3).

55. Prum 2018.

56. Keshishian 1979, 798–799.

57. Mirante 2006.

58. Khoo Thwe as quoted in Mirante 2006.

59. Quoted in Mydans 1996.

60. Quoted in Mydans 1996.

61. Theurer 2014, 51–67.

CHAPTER EIGHT. MEMBERSHIP & STATUS

1. Dokoupil 2008.

2. Nazaryan 2017; Ford 2017; Teitlell 2017.

3. Dizik 2014.

4. Fortes 1980, 2.

5. Mackrell 1986, 13–15.

6. Blackman 2015.

7. "Scarf, Votes for Women," National Museums Liverpool, www.liverpoolmuseums.org.uk/artifact/scarf-votes-women, accessed July 10, 2023.

8. Quotation from the journal *Votes for Women* (1908), cited by Fairhall 2006, 31.

9. Schmidt 2022.

10. Wickman 2012.

11. Stewart 2013.

12. Howe 1978.

13. Wickman 2012.

14. Kelly 2023.

15. Though commonly associated with Catholicism, the typical modern clerical collar was probably invented by a Scottish Presbyterian, Rev. Donald McLeod, at least according to reporting in the *Glasgow Herald*, December 6, 1894.

16. *BBC News* 2007.

17. Wynne-Jones 2007.

18. Watkinson 2016.

19. "Neckbeard," Know Your Meme, https://knowyourmeme.com/memes/neckbeard, accessed October 14, 2022.

20. Huber 1995, 145–148.

21. McCarter 2012, 34.

22. In a parallel strain of history, the association between red necks and defiant Appalachian people was reiterated in an early twentieth-century labor

movement. When the United Mine Workers began organizing coal miners in Appalachia, they adopted a red bandana worn around the neck as a symbol to unify their coalition of white, Black, and immigrant union members. Many union members proudly self-identified as "rednecks," but the term also became a pejorative slur used by mine operators and strike-breakers as a Red Scare tactic to (unfairly) link the striking miners to Communism (Huber 2006, 195).

23. Fisman and Sullivan 2016.

24. Fisman and Sullivan 2016.

25. Losos 2011, 22–23.

26. Losos 2011, 12–14.

27. Losos 2011, 173.

28. Recent genetic evidence indicates that females may have more possibility for mate choice than previously thought. Females may in fact choose between males in two adjacent territories or between the territorial male and interlopers.

29. Losos 2011, 297–301.

30. Losos 2011, 298.

31. *Neckclothitania* 1820.

32. Hart 1998, 41.

33. Wolfe 2002, 338.

34. Morrison 2015.

35. Young 2013.

36. Deihi 2015.

37. Vincent 2009, 18.

38. Plankensteiner 2007, 74–87.

39. Okpokunu, Agbontaen-Eghafona, and Ojo 2005, 160.

40. "Benin Memorial Head," Minneapolis Institute of Art, https://new.artsmia.org/programs/teachers-and-students/teaching-the-arts/artwork-in-focus/benin-momorial-head, accessed June 13, 2023.

41. Rohwer 1975, 593.

42. Rohwer 1985, 1325.

43. Ephron 2008.

44. Ephron 2008, 5.

45. Ephron 2008, 5.

46. Kamer and Pieper 2001, 123–127.

47. Ephron 2008, 7.

CHAPTER NINE. POWER & POLITICS

1. Zilczer 2001, 18–23.

2. Hyser and Downey 1987, 85.

3. Stein 1994.

4. Román-Palacios, Scholl, and Wiens 2019, 399.

5. Canids have longer necks than felids, probably because canids need their nose to reach near the ground for their olfactory-based lifestyle.

6. Fowler, Freedman, and Scannella 2009, 1.

7. McHenry et al. 2007, 16010.

8. Figueirido et al. 2018, 2360.

9. Brown 2014.

10. Recent analyses of saber-bearing predators indicate that species varied considerably in their bite strength. Some species may have had even weaker bites than the classic saber-toothed cat and relied even more on an especially strong neck. Others probably had bite strengths comparable to contemporary lions and killed their prey with a similar jaw-driven mechanism (Figueirido et al. 2018).

11. Sustaita, Rubega, and Farabaugh 2018.

12. Researchers have also speculated that the gargantuan dinosaur predator, *Tyrannosaurus rex*, also used its thick muscular neck in bite-and-shake predation. Their front limbs were short and probably useless in apprehending prey. Without talons or paws these 5,000-kilogram (11,000-pound) megapredators probably relied exclusively on their jaws and neck to kill, ironically, much like their 50-gram (1.7-ounce) living cousins, the shrikes (Snively et al. 2014, 290).

13. Fish et al. 2007, 2811.

14. Leviathan, the mythical sea serpent featured in the biblical book of Job, is sometimes equated to a crocodile (Revised Standard Version). Writers or translators of Job were also perhaps aware of how the necks of these terrifying beasts contributed to their ferociousness. The description of Leviathan in Job 41:22 says, "In his neck abides strength, and terror dances before him."

15. Dembitzer et al. 2022, [page 7].

16. Aghwan and Regenstein 2019, 111–121.

17. Vogel 2003, 209–255.

18. Livestock has also been used to power machines, such as grist mills.

19. Premi 1979.

20. Humans, both enslaved and free, have also borne yokes for transporting water and other cargo.

21. Bogucki 1993, 498.
22. Morenz and Kuhn 2022, 81.
23. *NPR* 2006.
24. Tyer 1992, 1026.
25. Yeivin and Rabinowitz 2007.
26. Tyer 1992, 1027.
27. In his thorough analysis of the passage in Gospel of Matthew, Matthew Mitchell (2016, 325) concludes that "easy" is not an accurate English translation of the original Greek text. He argues that "yoke" was almost certainly a metaphor for "commitment," but it is unclear whether this obligation is joyful or onerous. Mitchell believes "beneficial" may be the best translation.
28. Shuler 2014b.
29. Shuler 2014a.
30. Shuler 2014b, 22–24.
31. Shuler 2014b, 27–37.
32. Shuler 2014b, 39–48.
33. Steiner 2022, 333–336.
34. Shuler 2014b, 59–63.
35. Shuler 2014b, 60.
36. Shuler 2014b, 129–142.
37. Shuler 2014a.
38. For example, in January 2023, Iran executed two political protesters by hanging (*PBS* 2023).
39. Shuler 2014a.
40. Telford 2021.
41. Cho 2022.
42. Toomer, Johnson, and Byrd 1931, 27.
43. Frost 2021.
44. Quoted in Frost 2021.
45. Arasse 1989, 4.
46. Arasse 1989, 11.
47. White 2018.
48. White 2018.
49. Arasse 1989, 48–53.
50. White 2018.
51. *BBC* 2020.
52. Blake 2020.
53. Buchanan, Bull, and Patel 2020.

54. Biden 2021.
55. Pippin as quoted in Pagano 1945.

CHAPTER TEN. SHIELDS & SAINTS

1. Lundberg et al. 1994, 354.
2. Willmann and Bolmont 2012, 166–168.
3.. Scott Raymond, interview on November 2, 2022, Southington, CT.
4. Palomeque-del-Cerro et al. 2017, 49–52.
5. Raymond, interview.
6. Navarro and Karlins 2008.
7. Goldberg and Rosenthal 1986, 65.
8. Xydakis et al. 2005, 497–498.
9. Tong and Beirne 2013, 421.
10. Breeze et al. 2011, 1274–1276.
11. Chan et al. 2016, 255.
12. Pliny the Elder 1855.
13. Pliny the Elder 1855, 433.
14. Pliny the Elder 1855, 343.
15. *Croatia Week* 2017.
16. *The Reliquarian* 2014.
17. Bloch 2015.
18. McHenry and MacKeith 1966, 386–392.
19. Perez-Martinez, Riley, and Whiting 2020, 245.
20. Young and Kardong 2010, 1521.
21. Pianka and Pianka 1970, 90.
22. Anquetin, Tong, and Claude 2017, 1–8.

EPILOGUE

1. "What Does It Mean to Be Human? The Oldest Pottery," Smithsonian National Museum of Natural History, https://humanorigins.si.edu/evidence/behavior/carrying-storing/oldest-pottery, accessed August 8, 2023.
2. Hübner and Ufken 2023.
3. Hübner and Ufken 2023.
4. George Pearlman, interview with the author, August 2, 2023.
5. Buck 2022.
6. *Handmade* 2021.

BIBLIOGRAPHY

Achebe, Nwando. 2018. "Love, Courtship, and Marriage in Africa." In *A Companion to African History*, edited by William Worger, Charles Amble, and Nwando Achebe, 119–142. Hoboken, NJ: Wiley Blackwell.

Aghwan, Zeiad Amjad, and Joe Mac Regenstein. 2019. "Slaughter Practices of Different Faiths in Different Countries." *Journal of Animal Science and Technology* 61(3): 111–121.

Allen, Murray. 2014. "The New Whiplash." In *Musculoskeletal Pain Emanating from the Head and Neck*, edited by Murray Allen, 1–4. New York: Routledge.

Anderson, Jack. 1986. *Ballet and Modern Dance: A Concise History*. Trenton, NJ: Princeton Book Company.

Ang, Adam. 2022. "Smartphone-Based COVID-19 Detection Test from Australia Shows High Accuracy." *Mobile Health News*, March 22, 2022.

Anquetin, Jérémy, Haiyan Tong, and Julien Claude. 2017. "A Jurassic Stem Pleurodire Sheds Light on the Functional Origin of Neck Retraction in Turtles." *Scientific Reports* 7(1): 1–10.

Arasse, Daniel. 1989. *The Guillotine and the Terror*. New York: Viking Adult.

Aristotle. 1882. *On the Parts of Animals*. Edited by William Ogle. London: K. Paul, French & Company.

———. 2011. *Problems, Volume I: Books 1–19*. Edited by Robert Mayhew. Loeb Classical Library 316. Cambridge, MA: Harvard University Press.

Aung, Toe, and David Puts. 2020. "Voice Pitch: A Window into the Communication of Social Power." *Current Opinion in Psychology* 33: 154–161.

BBC. 2020. "George Floyd: What Happened in the Final Moments of His Life." July 16, 2020. www.bbc.com/news/world-us-canada-52861726. Accessed April 11, 2023.

BBC News. 2007. "Satanist Guilty of Vicar Killing." October 16, 2007. http://news.bbc.co.uk/2/hi/uk_news/wales/7047096.stm. Accessed June 14, 2023.

Belluck, Pam. 2021. "First Successful Trachea Transplant a Medical Milestone." *New York Times*, April 6, 2021.

Benjafield, Adam, Najib Ayas, Peter Eastwood, Raphael Heinzer, Mary Ip, Mary J. Morrell, Carlos Nunez, et al. 2019. "Estimation of the Global Prevalence and Burden of Obstructive Sleep Apnoea: A Literature-Based Analysis." *The Lancet: Respiratory Medicine* 7(8): 687–698.

Bergevin, Christopher, Chandan Narayan, Joy Williams, Natasha Mhatre, Jennifer K.E. Steeves, Joshua Bernstein, and Brad Story. 2020. "Overtone Focusing in Biphonic Tuvan Throat Singing." *eLife* 9: e50476.

Biden, Joseph. 2021. "Biden's Speech to Congress: Full Transcript." *New York Times*, September 7, 2021.

Biondi, Annachiara. 2022. "Neck's Best Thing: Beauty of the Choker Necklace." *Financial Times*, June 20, 2022.

Bitti, Federico. 2016. "Dystonia: Rewiring the Brain through Movement and Dance." TEDxNapoli, www.youtube.com/watch?v=DwkHK3rfKOo&ab_channel=TEDxTalks. Accessed July 12, 2022.

Blackman, Cally. 2015. "How the Suffragettes Used Fashion to Further the Cause." *The Guardian*, October 8, 2015.

Blake, Sam. 2020. "Why the George Floyd Protests Feel Different—Lots and Lots of Mobile Video." *dot.LA*, June 12, 2020. https://dot.la/george-floyd-video-2646171522.html?utm_campaign=post-teaser&utm_content=i87yytb3. Accessed June 15, 2023.

Bloch, Marc. 2015. *The Royal Touch: Sacred Monarchy and Scrofula in England and France*. New York: Routledge.

Boë, Louis-Jean, Thomas Sawallis, Joël Fagot, Pierre Badin, Guillaume Barbier, Guillaume Captier, Lucie Ménard, Jean-Louis Heim, and Jean-Luc Schwartz. 2019. "Which Way to the Dawn of Speech? Reanalyzing Half a Century of Debates and Data in Light of Speech Science." *Science Advances* 5(12): eaaw3916.

Bogucki, Peter. 1993. "Animal Traction and Household Economies in Neolithic Europe." *Antiquity* 67(256): 492–503.

Borkowska, Barbara, and Boguslaw Pawlowski. 2011. "Female Voice Frequency in the Context of Dominance and Attractiveness Perception." *Animal Behaviour* 82(1): 55–59.

Breeze, Major John, Celia Watson, Ian Horsfall, and Colonel Jon Clasper. 2011. "Comparing the Comfort and Potential Military Performance Restriction of Neck Collars from the Body Armor of Six Different Countries." *Military Medicine* 176(11): 1274–1277.

Broder, John. 2011. "E.P.A. Plans First Rules Ever on Perchlorate in Drinking Water." *New York Times*, February 2, 2011.

Brown, Jeffrey G. 2014. "Jaw Function in *Smilodon fatalis*: A Reevaluation of the Canine Shear-Bite and a Proposal for a New Forelimb-Powered Class 1 Lever Model." *PloS One* 9(10): e107456.

Bruno, Nicola, and Marco Bertamini. 2013. "Self-Portraits: Smartphones Reveal a Side Bias in Non-artists." *PloS One* 8: e55141.

Buchanan, Larry, Quoctrung Bull, and Jugal Patel. 2020. "Black Lives Matter May Be the Largest Movement in U.S. History." *New York Times*, July 3, 2020.

Buck, Louisa. 2022. "Magdalene Odundo Discusses Dancing with Clay ahead of Venice Biennale Exhibition." *The Art Newspaper*, March 28, 2022.

Button, Brian, Henry Goodell, Eyad Atieh, Yu-Cheng Chen, Robert Williams, Siddharth Shenoy, Elijah Lackey, et al. 2018. "Roles of Mucus Adhesion and Cohesion in Cough Clearance." *Proceedings of the National Academy of Sciences* 115(49): 12501–12506.

Carroll, Linda, Lena Holm, Robert Ferrari, Dejan Ozegovic, and J. David Cassidy. 2009. "Recovery in Whiplash-Associated Disorders: Do You Get What You Expect?" *Journal of Rheumatology* 36: 1063–1070.

Chan, Christie W. L., Janice J. Eng, Charles H. Tator, Andrei Krassioukov, and Spinal Cord Injury Research Evidence Team. 2016. "Epidemiology of Sport-Related Spinal Cord Injuries: A Systematic Review." *Journal of Spinal Cord Medicine* 39(3): 255–264.

Chang, Brian, Matthew Croson, Lorian Straker, Sean Gart, Carla Dove, John Gerwin, and Sunghwan Jung. 2016. "How Seabirds Plunge-Dive without Injuries." *Proceedings of the National Academy of Sciences* 113(43): 12006–12011.

Chen, Yuying, Ying Tang, Lawrence Vogel, and Michael DeVivo. 2013. "Causes of Spinal Cord Injury." *Topics in Spinal Cord Injury Rehabilitation* 19(1): 1–8.

Chen, Zhuo, and John Wiens. 2020. "The Origins of Acoustic Communication in Vertebrates." *Nature Communications* 11(1): 1–8.

Cho, Kelly. 2022. "Noose Found at Obama Presidential Construction Site in Chicago." *Washington Post*, November 10, 2022.

Chozick, Amy. 2023. "Liz Holmes Wants You to Forget about Elizabeth." *New York Times*, May 7, 2023.

Clayton, Martin, and Ronald Philo. 2010. *Leonardo da Vinci: The Mechanics of Man.* Los Angeles: Getty Publications.

Cler, Gabriel, Victoria McKenna, Kimberly Dahl, and Cara Stepp. 2020. "Longitudinal Case Study of Transgender Voice Changes under Testosterone Hormone Therapy." *Journal of Voice* 34(5): 748–762.

Colapinto, John. 2022. *This Is Your Voice.* New York: Simon and Schuster.

Cole, Jonathan. 1995. *Pride and a Daily Marathon.* Cambridge, MA: MIT Press.

Collis, Helen, 2022. "Romania to Issue Iodine Tablets as Russian War Continues in Neighboring Ukraine." *Politico*, April 3, 2022.

Conyn, Cornelius. 1953. *Three Centuries of Ballet.* Houston, TX: Elsevier Press.

Costa, Marco, Marzia Menzani, and Pio Enrico Ricci Bitti. 2001. "Head Canting in Paintings: An Historical Study." *Journal of Nonverbal Behavior* 25(1): 63–73.

Couderq, Stephan, Michelle Leemans, and Jean-Baptiste Fini. 2020. "Testing for Thyroid Hormone Disruptors, a Review of Non-mammalian In Vivo Models." *Molecular and Cellular Endocrinology* 508: 110779.

Cowell, Alan. 2008. "Europeans Announce Pioneering Stem Cell Surgery." *New York Times*, November 19, 2008.

Crawford, John. 1964. "Living without a Balancing Mechanism." *British Journal of Ophthalmology* 48(7): 357–360.

Croatia Week. 2017. "Dubrovnik Celebrates 1,045th Anniversary of Its Patron Saint." *Croatia Week*, February 3, 2017. www.croatiaweek.com/dubrovnik-celebrates-1045th-anniversary-of-its-patron-saint/. Accessed February 12, 2023.

Crockford, Susan. 2009. "Evolutionary Roots of Iodine and Thyroid Hormones in Cell–Cell Signaling." *Integrative and Comparative Biology* 49(2): 155–166.

Daeschler, Edward B., Neil H. Shubin, and Farish A. Jenkins Jr. 2006. "A Devonian Tetrapod-Like Fish and the Evolution of the Tetrapod Body Plan." *Nature* 440(7085): 757–763.

Damas, Aline. 2018 "Revisiting the Female Gazes in Manet's 'Olympia.'" *The Harvard Crimson*, April 24, 2018.

Darwin, Charles. 1871. *The Descent of Man, and Selection in Relation to Sex.* London: John Murray, Albemarle Street.

Dave, Bharat, Ajay Krishnan, Ravi Ranjan Rai, Devanand Degulmadi, and Shivanand Mayi. 2021. "The Effect of Head Loading on Cervical Spine in Manual Laborers." *Asian Spine Journal* 15(1): 17–22.

De Boer, Bart. 2012. "Loss of Air Sacs Improved Hominin Speech Abilities." *Journal of Human Evolution* 62(1): 1–6.

Deihi, Nancy. 2015. "A Scarf Can Mean Many Things—but Above All, Prestige." *The Conversation*, May 15, 2015.

de Kok-Mercado, Fabian, Michael Habib, Tim Phelps, Lydia Gregg, and Philippe Gailloud. 2013. "Adaptations of the Owl's Cervical and Cephalic Arteries in Relation to Extreme Neck Rotation." *Science* 339(6119): 514–514.

Dembitzer, Jacob, Ran Barkai, Miki Ben-Dor, and Shai Meiri. 2022. "Levantine Overkill: 1.5 Million Years of Hunting Down the Body Size Distribution." *Quaternary Science Reviews* 276: 107316 (online).

Dillon, Cheryl. 2017. "Super Model." Patient Stories. Dystonia Medical Research Foundation Canada. https://dystoniacanada.org/sites/dystoniacanada.org/files/2017-04/If-I-were-a-Supermodel%20by%20Cheryl%20Dillon.pdf. Accessed July 13, 2023.

Dizik, Alina. 2014. "What the Color of Your Tie Says about You." *BBC*, August 31, 2014.

Dokoupil, Tony. 2008. "Candidates' Neckties: What Their Knots Mean." *Newsweek*, October 13, 2008.

Dudley, Robert, and A. Stanley Rand. 1991. "Sound Production and Vocal Sac Inflation in the Túngara Frog, *Physalaemus pustulosus* (Leptodactylidae)." *Copeia*: 460–470.

Elemans, Coen P.H., Andrew F. Mead, Lawrence C. Rome, and Franz Goller. 2008. "Superfast Vocal Muscles Control Song Production in Songbirds." *PloS One* 3(7): e2581 (online).

Elflein, John. N.d. "Number of Choking Deaths in the US, 1945–2022." Statistica. www.statista.com/statistics/527321/deaths-due-to-choking-in-the-us/. Accessed March 21, 2023.

Elliott, James, Peter McMenamin, and David Walton. 2016. "Chronic Whiplash: Is It Really a Medical Mystery?" *PT Think Tank*. https://ptthinktank.com/2016/10/19/chronic-whiplash-is-it-really-a-medical-mystery/. Accessed June 16, 2023.

Ephron, Nora. 2008. *I Feel Bad about My Neck*. New York: Random House.

Ernstbrunner, Lukas, Armin Runer, Paul Siegert, Matthäus Ernstbrunner, Johannes Becker, Thomas Freude, Herbert Resch, and Philipp Moroder. 2017. "A Prospective Analysis of Injury Rates, Patterns and Causes in Cliff and Splash Diving." *Injury* 48(10): 2125–2131.

Fairhall, David 2006. *Common Ground: The Story of Greenham*. London: Bloomsbury Publishing.

Feinberg, David R., Benedict C. Jones, and Marie M. Armstrong. 2018. "Sensory Exploitation, Sexual Dimorphism, and Human Voice Pitch." *Trends in Ecology & Evolution* 33(12): 901–903.

———. 2019. "No Evidence That Men's Voice Pitch Signals Formidability." *Trends in Ecology & Evolution* 34(3): 190–192.

Ferrari, Robert. 2006. *The Whiplash Encyclopedia: The Facts and Myths of Whiplash.* Sudbury, MA: Jones and Bartlett Publishers.

Ferrari, Robert, Constantine Constantoyannis, and Nikolas Papadakis. 2001. "Cross-Cultural Study of Symptom Expectation Following Minor Head Injury in Canada and Greece." *Clinical Neurology and Neurosurgery* 103(4): 254–259.

Fiebert, Ira, Fran Kistner, Christine Gissendanner, and Christopher DaSilva. 2021. "Text Neck: An Adverse Postural Phenomenon." *Work* 69: 1261–1270.

Figueirido, Borja, Stephan Lautenschlager, Alejandro Pérez-Ramos, and Blaire Van Valkenburgh. 2018. "Distinct Predatory Behaviors in Scimitar- and Dirk-Toothed Sabertooth Cats." *Current Biology* 28: 3260–3266.

Fish, Frank E., Sandra Bostic, Anthony Nicastro, and John Beneski. 2007. "Death Roll of the Alligator: Mechanics of Twist Feeding in Water." *Journal of Experimental Biology* 210: 2811–2818.

Fisman, Raymond, and Tim Sullivan. 2016. "The Case for Neck Tattoos according to Economists." *The Atlantic*, June 13, 2016.

Fitch, W. Tecumseh. 1999. "Acoustic Exaggeration of Size in Birds via Tracheal Elongation: Comparative and Theoretical Analyses." *Journal of Zoology* 248(1): 31–48.

———. 2018. "The Biology and Evolution of Speech: A Comparative Analysis." *Annual Review of Linguistics* 4: 255–279.

Fitch, W. Tecumseh, Bart De Boer, Neil Mathur, and Asif A. Ghazanfar. 2016. "Monkey Vocal Tracts Are Speech-Ready." *Science Advances* 2(12): e1600723.

Ford, Richard Thompson. 2017. "The Ties That Blind." *New York Times*, February 10, 2017.

Fortes, Meyer. 1980. "The Necktie." *Cambridge Anthropology* 6: 1–9.

Fountain, Henry. 2012. "Synthetic Windpipe Is Used to Replace Cancerous One." *New York Times*, January 12, 2012.

Fowler, Denver, Elizabeth Freedman, and John Scannella. 2009. "Predatory Functional Morphology in Raptors: Interdigital Variation in Talon Size Is Related to Prey Restraint and Immobilisation Technique." *PloS One* 4(11): e7999.

Fraccaro, Paul, Benedict Jones, Jovana Vukovic, Finlay Smith, Christopher Watkins, David Feinberg, Anthony Little, and Lisa Debruine. 2011. "Experimental Evidence That Women Speak in a Higher Voice Pitch to Men They Find Attractive." *Journal of Evolutionary Psychology* 9(1): 57–67.

Friedman, Lisa. 2022. "E.P.A. Decides against Limiting Perchlorate in Drinking Water." *New York Times*, March 31, 2022.

Frost, Natasha. 2021. "He Calls the Tie a Colonial Noose. Now, Parliament Says It's No Longer Mandatory." *New York Times*, February 10, 2021.

Gans, Carl. 1992. "Why Develop a Neck?" In *The Head-Neck Sensory Motor System*, edited by A. Berthoz, W. Graf, and P. Vidal, 17–22. New York: Oxford University Press.

Gil, Kelsey, A. Wayne Vogl, and Robert Shadwick. 2022. "Anatomical Mechanism for Protecting the Airway in the Largest Animals on Earth." *Current Biology* 32(4): 898–903.

Goldberg, Shelly, and Robert Rosenthal. 1986. "Self-Touching Behavior in the Job Interview: Antecedents and Consequences." *Journal of Nonverbal Behavior* 10(1): 65–80.

Goldbogen, Jeremy. 2010. "The Ultimate Mouthful: Lunge Feeding in Rorqual Whales." *American Scientist* 98(2): 124–131.

Grajal, Alejandro, Stuart Strahl, Rodrigo Parra, Maria Gloria Dominguez, and Alfredo Neher. 1989. "Foregut Fermentation in the Hoatzin, a Neotropical Leaf-Eating Bird." *Science* 245(4923): 1236–1238.

Grant, Ian. 1989. "Nyanza Watering-Place: The Remarkable Story of the SS *William Mackinnon*." *Review of Scottish Culture* 5: 63–78.

Gray, Henry. 1988. *Gray's Anatomy*. Gramercy.

Gross, Charles. 1998. "Galen and the Squealing Pig." *The Neuroscientist* 4(3): 216–221.

Hale, Conor. 2022. "Pfizer to Drop $74M for COVID Cough-Screening Smartphone App Developer." *Fierce Biotech*, April 11, 2022. www.fiercebiotech.com/medtech/pfizer-drop-74m-covid-cough-screening-smartphone-app-developer.

Handmade. 2021. "How a Master Potter Makes Giant Kimchi Pots Using the Traditional Method." *Handmade*. www.youtube.com/watch?v=QlwnBy16WoE. Accessed October 2, 2023.

Haneline, Michael. 2009. "The Notion of a 'Whiplash Culture': A Review of the Evidence." *Journal of Chiropractic Medicine* 8(3): 119–124.

Hannan, S. Alam. 2006. "The Magnificent Seven: A History of Modern Thyroid Surgery." *International Journal of Surgery* 4(3): 187–191.

Hansraj, Kenneth. 2014. "Assessment of Stresses in the Cervical Spine Caused by Posture and Position of the Head." *Surgical Technology International* 25(25): 277–279.

Hart, Avril. 1998. *Ties*. New York: Costume & Fashion Press.

Heglund, Norman, Patrick Willems, Massimo Penta, and Giovanni Cavagna. 1995. "Energy-Saving Gait Mechanics with Head-Supported Loads." *Nature* 375(6526): 52–54.

Held, Lewis, Jr. 2009. *Quirks of Human Anatomy: An Evo-Devo Look at the Human Body*. Cambridge, UK: Cambridge University Press.

Herbst, Christian, Tamara Prigge, Maxime Garcia, Vit Hampala, Riccardo Hofer, Gerald Weissengruber, Jan Svec, and W. Tecumseh Fitch. 2023. "Domestic Cat Larynges Can Produce Purring Frequencies without Neural Input." *Current Biology* 33(21): 4727–4732.

Herbst, Christian, Angela Stoeger, Roland Frey, Jörg Lohscheller, Ingo Titze, Michaela Gumpenberger, and W. Tecumseh Fitch. 2012. "How Low Can You Go? Physical Production Mechanism of Elephant Infrasonic Vocalizations." *Science* 337(6094): 595–599.

Holm, Lena, Linda Carroll, J. David Cassidy, Eva Skillgate, and Anders Ahlbom. 2008. "Expectations for Recovery Important in the Prognosis of Whiplash Injuries." *PLoS Medicine* 5(5): e105.

Hovet, Jason. 2022. "Putin's Nuclear Comments Lead to Rush for Iodine in Central Europe." Reuters, March 2, 2022. www.reuters.com/world/europe/putins-nuclear-comments-lead-rush-iodine-central-europe-2022-03-02/. Accessed July 23, 2023.

"How." N.d. "How Animals Holler." University of Utah. https://phys.org/news/2018-05-animals-holler.html. Accessed July 2, 2023.

Howe, Louise Kapp. 1978. *Pink Collar Workers*. New York: Avon.

Howes, Patricia, Jean Callaghan, Pamela Davis, Dianna Kenny, and William Thorpe. 2004. "The Relationship between Measured Vibrato Characteristics and Perception in Western Operatic Singing." *Journal of Voice* 18(2): 216–230.

Huber, Patrick. 1995. "A Short History of Redneck: The Fashioning of a Southern White Masculine Identity." *Southern Cultures* 1(2): 145–166.

———. 2006. "Red Necks and Red Bandanas: Appalachian Coal Miners and the Coloring of Union Identity, 1912–1936." *Western Folklore* 65: 195–210.

Hübner, Ronald, and Emily Sophie Ufken. 2023. "On the Beauty of Vases: Birkhoff's Aesthetic Measure versus Hogarth's Line of Beauty." *Frontiers in Psychology* 14: 1114793 (online).

Hughes, Susan M., and David A. Puts. 2021. "Vocal Modulation in Human Mating and Competition." *Philosophical Transactions of the Royal Society B* 376(1840): 20200388 (online).

Hyser, Raymond, and Dennis Downey. 1987. "'A Crooked Death': Coatesville, Pennsylvania and the Lynching of Zachariah Walker." *Pennsylvania History: A Journal of Mid-Atlantic Studies* 54(2): 85–102.

Jakobsen, Lasse, Jakob Christensen-Dalsgaard, Peter Møller Juhl, and Coen Elemans. 2021. "How Loud Can You Go? Physical and Physiological Constraints to Producing High Sound Pressures in Animal Vocalizations." *Frontiers in Ecology and Evolution* 9: 657254 (online).

James, Logan, A. Leonie Baier, Rachel Page, Paul Clements, Kimberly Hunter, Ryan Taylor, and Michael Ryan. 2022. "Cross-Modal Facilitation of Auditory Discrimination in a Frog." *Biology Letters* 18(6): 20220098 (online).

Jarvis, Erich. 2019. "Evolution of Vocal Learning and Spoken Language." *Science* 366 (6461): 50–54.

Johansson, Ellen, Louise Andersson, Jessica Örnros, Therese Carlsson, Camilla Ingeson-Carlsson, Shawn Liang, Jakob Dahlberg, et al. 2015. "Revising the Embryonic Origin of Thyroid C Cells in Mice and Humans." *Development* 142(20): 3519–3528.

Johns Hopkins Medicine. 2020. "Something in the Water: Pollutant May Be More Hazardous Than Previously Thought." *ScienceDaily*, June. www.sciencedaily.com/releases/2020/06/200605121514.htm. Accessed June 6, 2022.

Kamer, Frank, and Patrick G. Pieper. 2001. "Surgical Treatment of the Aging Neck." *Facial Plastic Surgery* 17: 123–128.

Kelly, Lora. 2023. "Wanted: 'New Collar Workers.'" *New York Times*, December 29, 2023.

Keshishian, John. 1979. "Anatomy of a Burmese Beauty Secret." *National Geographic* 155(6): 798–801.

Kimani, James Kirumbi. 1987. "Structural Organization of the Vertebral Artery in the Giraffe (*Giraffa camelopardalis*)." *The Anatomical Record* 217(3): 256–262.

Kingsley, Evan, Chad Eliason, Tobias Riede, Zhiheng Li, Tom Hiscock, Michael Farnsworth, Scott Thomson, Franz Goller, Clifford Tabin, and Julia Clarke. 2018. "Identity and Novelty in the Avian Syrinx." *Proceedings of the National Academy of Sciences* 115(41): 10209–10217.

Klofstad, Casey, Stephen Nowicki, and Rindy Anderson. 2016. "How Voice Pitch Influences Our Choice of Leaders: When Candidates Speak, Their Vocal Characteristics—as Well as Their Words—Influence Voters' Attitudes toward Them." *American Scientist* 104(5): 282–288.

Krings, Markus, John Nyakatura, Mark Boumans, Martin Fischer, and Hermann Wagner. 2017. "Barn Owls Maximize Head Rotations by a Combination of Yawing and Rolling in Functionally Diverse Regions of the Neck." *Journal of Anatomy* 231(1): 12–22.

Kristoff, Nicholas. 2008. "Raising the World's I.Q." *New York Times*, December 4, 2008.

Laguarta, Jordi, Ferran Hueto, and Brian Subirana. 2020. "COVID-19 Artificial Intelligence Diagnosis Using Only Cough Recordings." *IEEE Open Journal of Engineering in Medicine and Biology* 1: 275–281.

Lee, Chen-Hsen, and Jen-Hwey Chiu. 2021. "Goiter Disease in Traditional Chinese Medicine: Modern Insight into Ancient Wisdom." *Journal of the Chinese Medical Association* 84(6): 577–579.

Leonardo da Vinci. 1952. *Leonardo da Vinci on the Human Body*. Edited by C. O'Malley and J. Saunders. New York: H. Schuman.

Leung, Angela, Lewis Braverman, and Elizabeth Pearce. 2012. "History of US Iodine Fortification and Supplementation." *Nutrients* 4(11): 1740–1746.

Lieberman, Daniel. 2011. *The Evolution of the Human Head*. Cambridge, MA: Harvard University Press.

Little, Alexander, and Frank Seebacher. 2014. "The Evolution of Endothermy Is Explained by Thyroid Hormone–Mediated Responses to Cold in Early Vertebrates." *Journal of Experimental Biology* 217(10): 1642–1648.

Liu, Chang, Jianbo Gao, Xinxin Cui, Zhipeng Li, Lei Chen, Yuan Yuan, Yaolei Zhang, et al. 2021. "A Towering Genome: Experimentally Validated Adaptations to High Blood Pressure and Extreme Stature in the Giraffe." *Science Advances* 7(12): eabe9459.

Lloyd, Ray, Bridget Parr, Simeon Davies, and Carlton Cooke. 2010. "Subjective Perceptions of Load Carriage on the Head and Back in Xhosa Women." *Applied Ergonomics* 41(4): 522–529.

Los Angeles Times. 1927. "Dancer Dies from Fall; Isadora Duncan Meets Fate." *Los Angeles Times*, September 15, 1927.

Losos, Jonathan B. 2011. *Lizards in an Evolutionary Tree: Ecology and Adaptive Radiation of Anoles*. Berkeley: University of California Press.

Louchart, Antoine, and Laurent Viriot. 2011. "From Snout to Beak: The Loss of Teeth in Birds." *Trends in Ecology & Evolution* 26(12): 663–673.

Lundberg, Ulf, Roland Kadefors, Bo Melin, Gunnar Palmerud, Peter Hassmén, Margareta Engström, and Ingela Elfsberg Dohns. 1994. "Psychophysiological Stress and EMG Activity of the Trapezius Muscle." *International Journal of Behavioral Medicine* 1(4): 354–370.

Lydiatt, Daniel, and Gregory Bucher. 2011. "Historical Vignettes of the Thyroid Gland." *Clinical Anatomy* 24(1): 1–9.

MacDonell, Nancy. 2022. "The Surprisingly Dark History of the Choker Necklace." *Wall Street Journal*, August 24, 2022.

MacIver, Malcolm A., and Barbara L. Finlay. 2022. "The Neuroecology of the Water-to-Land Transition and the Evolution of the Vertebrate Brain." *Philosophical Transactions of the Royal Society B* 377(1844): 20200523 (online).

Mackrell, Alice. 1986. *Shawls, Stoles and Scarves*. London: Batsford.

Markova, Diana, Louis Richer, Melissa Pangelinan, Deborah Schwartz, Gabriel Leonard, Michel Perron, G. Bruce Pike, et al. 2016. "Age- and Sex-Related Variations in Vocal-Tract Morphology and Voice Acoustics during Adolescence." *Hormones and Behavior* 81: 84–96.

McCarter, William Matthew. 2012. *Homo Redneckus: On Being Not White in America*. New York: Algora Publishing.

McHenry, Colin, Stephen Wroe, Philip Clausen, Karen Moreno, and Eleanor Cunningham. 2007. "Supermodeled Sabercat: Predatory Behavior in *Smilodon fatalis* Revealed by High-Resolution 3D Computer Simulation." *Proceedings of the National Academy of Sciences* 104(41): 16010–16015.

McHenry, Lawrence, and Ronald MacKeith. 1966. "Samuel Johnson's Childhood Illnesses and the King's Evil." *Medical History* 10(4): 386–399.

Mirante, Edith. 2006. "The Dragon Mothers Polish Their Metal Coils." *Guernica*, September 28, 2006.

Mitchell, Matthew. 2016. "The Yoke Is Easy, but What of Its Meaning? A Methodological Reflection Masquerading as a Philological Discussion of Matthew 11:30." *Journal of Biblical Literature* 135(2): 321–340.

Moore, Monica. 1985. "Nonverbal Courtship Patterns in Women: Context and Consequences." *Ethology and Sociobiology* 6(4): 237–247.

Morenz, Ludwig, and Robert Kuhn. 2022. "Tax Coercion as a Real and Metaphorical YOKE: On the Earliest State Administrative Practices Reflected in Ancient Egyptian Writing and Images around 3000 BC." In *Slavery and Other Forms of Strong Asymmetrical Dependencies*, edited by Jeannine Bischoff and Stephan Conermann, 73–89. Berlin: DeGruyter.

Morimoto, Takaaki, Jun-ichiro Enmi, Yorito Hattori, Satoshi Iguchi, Satoshi Saito, Kouji Harada, Hiroko Okuda, et al. 2018. "Dysregulation of RNF213 Promotes Cerebral Hypoperfusion." *Scientific Reports* 8(1): 1–9.

Morrison, Lennox. 2015. "Is This the New Power Symbol for Women?" *BBC*, March 19, 2015.

Mughal, Bilal, Jean-Baptiste Fini, and Barbara Demeneix. 2018. "Thyroid-Disrupting Chemicals and Brain Development: An Update." *Endocrine Connections* 7(4): R160–R186.

Mydans, Seth. 1996. "New Thai Tourist Sight: Burmese 'Giraffe Women.'" *New York Times,* October 19, 1996.

Natterson-Horowitz, Barbara, Basil Baccouche, Jennifer Mary Head, Tejas Shivkumar, Mads Frost Bertelsen, Christian Aalkjær, Morten Smerup, Olujimi Ajijola, Joseph Hadaya, and Tobias Wang. 2021. "Did Giraffe Cardiovascular Evolution Solve the Problem of Heart Failure with Preserved Ejection Fraction?" *Evolution, Medicine, and Public Health* 9(1): 248–255.

Nature. 2016. "Macchiarini Scandal Is a Valuable Lesson for the Karolinska Institute." *Nature* 537(137). https://doi.org/10.1038/537137a.

Navarro, Joe, and Marvin Karlins. 2008. *What Every Body Is Saying.* New York: HarperCollins Publishers.

Nazaryan, Alexander. 2017. "Trump's Codpiece: What's with the Long Ties?" *Newsweek*, February 13, 2017.

Neckclothitania. 1820. London: J. J. Stockdale.

Nicholls, Michael, Danielle Clode, Stephen Wood, and Amanda Wood. 1999. "Laterality of Expression in Portraiture: Putting Your Best Cheek Forward." *Proceedings of the Royal Society of London, Series B: Biological Sciences* 266(1428): 1517–1522.

Nishimura, Takeshi, Isao Tokuda, Shigehiro Miyachi, Jacob Dunn, Christian Herbst, Kazuyoshi Ishimura, Akihisa Kaneko, et al. 2022. "Evolutionary Loss of Complexity in Human Vocal Anatomy as an Adaptation for Speech." *Science* 377(6607): 760–763.

Nooshin, Laudan. 1998. "The Song of the Nightingale: Processes of Improvisation in *Dastgāh Segāh* (Iranian classical music)." *British Journal of Ethnomusicology* 7(1): 69–116.

NPR. 2006. "Up Close and Personal with the Albatross." *NPR*, November 24, 2006.

———. 2011. "From Meryl to Margaret: Becoming 'The Iron Lady.'" *NPR*, December 19, 2011.

Okabe, Masataka, and Anthony Graham. 2004. "The Origin of the Parathyroid Gland." *Proceedings of the National Academy of Sciences* 101(51): 17716–17719.

Okpokunu, Edoja, Kokunre Agbontaen-Eghafona, and Pat Ojo. 2005. "Benin Dressing in Contemporary Nigeria: Social Change and the Crisis of Cultural Identity." *African Identities* 3(2): 155–170.

Pagano, Grace. 1945. *Contemporary American Painting. The Encyclopædia Britannica Collection.* New York: Duell, Sloan and Pearce.

Paley, William. 1854. *Natural Theology: or, Evidences of the Existence and Attributes of the Deity, Collected from the Appearances of Nature.* Boston: Gould and Lincoln.

Palomeque-del-Cerro, Luis, Luis Arraez-Aybar, Cleofas Rodriguez-Blanco, Rafael Guzman-Garcia, Mar Menendez-Aparicio, and Angel Oliva-Pascual-Vaca. 2017. "A Systematic Review of the Soft-Tissue Connections between Neck Muscles and Dura Mater: The Myodural Bridge." *Spine* 42(1): 49–54.

Pavela Banai, Irena. 2017. "Voice in Different Phases of the Menstrual Cycle among Naturally Cycling Women and Users of Hormonal Contraceptives." *PLoS One* 12(8): e0183462.

PBS. 2023. "Iran Executes 2 More Men Detained during Nationwide Protests." *PBS*, January 7, 2023.

Pentreath, Rosie. 2021. "Thirteen Pieces of Classical Music Inspired by Birdsong." *Discover Music*, August 23, 2021. www.classicfm.com/discover-music/classical-music-inspired-by-birdsong/. Accessed March 13, 2023.

Perez-Martinez, Christian, Julia Riley, and Martin Whiting. 2020. "Uncovering the Function of an Enigmatic Display: Antipredator Behaviour in the Iconic Australian Frillneck Lizard." *Biological Journal of the Linnean Society* 129(2): 425–438.

Pianka, Eric, and Helen D. Pianka. 1970. "The Ecology of *Moloch horridus* (Lacertilia: Agamidae) in Western Australia." *Copeia*: 90–103.

Pipitone, Nathan, and Gordon Gallup Jr. 2008. "Women's Voice Attractiveness Varies across the Menstrual Cycle." *Evolution and Human Behavior* 29(4): 268–274.

Pisanski, Katarzyna, and David Reby. 2021. "Efficacy in Deceptive Vocal Exaggeration of Human Body Size." *Nature Communications* 12(1): 1–9.

Pisanski, Katarzyna, Paul Fraccaro, Cara Tigue, Jillian O'Connor, Susanne Röder, Paul Andrews, Bernhard Fink, Lisa DeBruine, Benedict Jones, and David Feinberg. 2014. "Vocal Indicators of Body Size in Men and Women: A Meta-analysis." *Animal Behaviour* 95: 89–99.

Pisanski, Katarzyna, Anna Oleszkiewicz, Justyna Plachetka, Marzena Gmiterek, and David Reby. 2018. "Voice Pitch Modulation in Human Mate Choice." *Proceedings of the Royal Society B* 285(1893): 20181634 (online).

Plankensteiner, Barbara. 2007. "Benin—Kings and Rituals: Court Arts from Nigeria." *African Arts* 40(4): 74–87.

Plato. 2013. *Republic.* Edited by Clair Emlyn-Jones and William Preddy, vol. 5. Cambridge, MA: Harvard University Press.

Plato. 2009. *Timaeus and Critias.* Oxford: Oxford University Press. ProQuest Ebook Central.

Pliny the Elder. 1855. *The Natural History, Book 28.* Translated by John Bostock, M.D., F.R.S. H.T. Riley, Esq., B.A. London: Taylor and Francis.

Podos, Jeffrey, and Mario Cohn-Haft. 2019. "Extremely Loud Mating Songs at Close Range in White Bellbirds." *Current Biology* 29(20): R1068–R1069.

Premi, S.C.L. 1979. "Performance of Bullocks in Varying Conditions of Load and Climate." Master's thesis, Asian Institute of Technology, Bangkok. Cited in *Mechanics of Pre-industrial Technology: An Introduction to the Mechanics of Ancient and Traditional Material Culture,* by Brian Cotterell and Johan Kamminga. Cambridge, UK: Cambridge University Press, 1992.

Preston-Whyte, Eleanor, and Jean Morris. 1994. *Speaking with Beads: Zulu Arts from Southern Africa.* London: Thames and Hudson.

Prum, Richard. 2018. *The Evolution of Beauty: How Darwin's Forgotten Theory of Mate Choice Shapes the Animal World—and Us.* New York: Anchor.

Puts, David. 2005. "Mating Context and Menstrual Phase Affect Women's Preferences for Male Voice Pitch." *Evolution and Human Behavior* 26(5): 388–397.

Puts, David, and Toe Aung. 2019. "Does Men's Voice Pitch Signal Formidability? A Reply to Feinberg et al." *Trends in Ecology & Evolution* 34(3): 189–190.

Puts, David, Alexander Hill, Drew Bailey, Robert Walker, Drew Rendall, John Wheatley, Lisa Welling, et al. 2016. "Sexual Selection on Male Vocal Fundamental Frequency in Humans and Other Anthropoids." *Proceedings of the Royal Society B: Biological Sciences* 283(1829): 20152830 (online).

Ramya, Rachna. 2019. *Kathak, the Dance of Storytellers.* New Delhi: Niyogi Books, 2019.

The Reliquarian. 2014. "Protector of Dubrovnik and Patron Saint of Throat Illnesses." *The Reliquarian,* May 19, 2014. https://reliquarian.com/2014/05/19/saint-blaise-protector-of-dubrovnik-and-patron-saint-of-throat-illnesses/.

Renninger, Lee Ann, T. Joel Wade, and Karl Grammer. 2004. "Getting That Female Glance: Patterns and Consequences of Male Nonverbal Behavior in Courtship Contexts." *Evolution and Human Behavior* 25(6): 416–431.

Riede, Tobias, Scott Thomson, Ingo Titze, and Franz Goller. 2019. "The Evolution of the Syrinx: An Acoustic Theory." *PloS Biology* 17(2): e2006507 (online).

Riede, Tobias, Isao Tokuda, Jacob Munger, and Scott Thomson. 2008. "Mammalian Laryngeal Air Sacs Add Variability to the Vocal Tract Impedance: Physical and Computational Modeling." *Journal of the Acoustical Society of America* 124(1): 634–647.

Rockel, Stephen J. 2006. *Carriers of Culture: Labor on the Road in Nineteenth-Century East Africa.* Westport, CT: Praeger.

Rohwer, Sievert. 1975. "The Social Significance of Avian Winter Plumage Variability." *Evolution*: 29(4):593–610.

———. 1985. "Dyed Birds Achieve Higher Social Status Than Controls in Harris' Sparrows." *Animal Behaviour* 33(4): 1325–1331.

Román-Palacios, Cristian, Joshua Scholl, and John Wiens. 2019. "Evolution of Diet across the Animal Tree of Life." *Evolution Letters* 3(4): 339–347.

Rozsa, Matthew. 2021. "The Human Neck Is a Mistake of Evolution." *Salon*, October 12, 2021. www.salon.com/2021/10/12/the-human-neck-is-an-evolutionary-mistake/. Accessed March 13, 2022.

Sadhra, Makita, H. Samaratunga, H. S. Ahmed, and Laura Tonks. 2015. "The Draining of a Lifetime." *Physics Special Topics* 14(1).

Sasaki, Clarence. 2006. "Anatomy and Development and Physiology of the Larynx." *GI Motility Online.* www.nature.com/gimo/contents/pt1/full/gimo7.html, doi:10.1038/gimo7. Accessed April 4, 2023.

Scally, Aylwyn, Julien Dutheil, LaDeana Hillier, Gregory Jordan, Ian Goodhead, Javier Herrero, Asger Hobolth, et al. 2012. "Insights into Hominid Evolution from the Gorilla Genome Sequence." *Nature* 483(7388): 169–175.

Schmidt, Samantha. 2022. "How Green Became the Color of Abortion Rights." *Washington Post*, July 3, 2022.

Schneider, Leonid. 2020. "Paolo Macchiarini Indicted for Aggravated Assault in Sweden." *For Better Science*, September 30, 2020. https://forbetterscience.com/2020/09/30/paolo-macchiarini-indicted-for-aggravated-assault-in-sweden/. Accessed September 9, 2023.

Schneider, Tobias M., and Claus-Christian Carbon. 2017. "Taking the Perfect Selfie: Investigating the Impact of Perspective on the Perception of Higher Cognitive Variables." *Frontiers in Psychology* 8: 244419 (online).

Schoeman, Stan. 1983. "Eloquent Beads: The Semantics of a Zulu Art Form." *Africa Insight* 13(2): 147–152.

Seashore, Carl. 1931. "The Natural History of the Vibrato." *Proceedings of the National Academy of Sciences* 17(12): 623–626.

Seashore, Carl, and Milton Metfessel. 1925. "Deviation from the Regular as an Art Principle." *Proceedings of the National Academy of Sciences* 11: 538–542.

Sedgewick, Jennifer, Meghan Flath, and Lorin Elias. 2017. "Presenting Your Best Self(ie): The Influence of Gender on Vertical Orientation of Selfies on Tinder." *Frontiers in Psychology* 8: 204804 (online).

Seymour, Roger, Vanya Bosiocic, and Edward Snelling. 2016. "Fossil Skulls Reveal That Blood Flow Rate to the Brain Increased Faster Than Brain Volume during Human Evolution." *Royal Society Open Science* 3(8): 160305.

Sharker, Saberul, Sean Holekamp, Mohammad Mansoor, Frank Fish, and Tadd Truscott. 2019. "Water Entry Impact Dynamics of Diving Birds." *Bioinspiration & Biomimetics* 14(5): 056013.

Shubin, Neil, Edward Daeschler, and Farish Jenkins. 2015. "Origin of the Tetrapod Neck and Shoulder." In *Great Transformations in Vertebrate Evolution*, edited by Kenneth Dial, Neil Shubin, and Elizabeth Brainerd, 63–76. Chicago: University of Chicago Press.

Shuler, Jack 2014a. "The Ominous Symbolism of the Noose." *Los Angeles Times*, October 27, 2014.

———. 2014b. *The Thirteenth Turn: A History of the Noose.* New York: Public Affairs.

Simmons, Robert, and Res Altwegg. 2010. "Necks-for-Sex or Competing Browsers? A Critique of Ideas on the Evolution of Giraffes." *Journal of Zoology* 282(1): 6–12.

Simmons, Robert, and Lue Scheepers. 1996. "Winning by a Neck: Sexual Selection in the Evolution of Giraffes." *The American Naturalist* 148(5): 771–786.

Sinervo, Barry, and Curt Lively. 1996. "The Rock-Paper-Scissors Game and the Evolution of Alternative Male Strategies." *Nature* 380(6571): 240–243.

"Singer." N.d. "Singer Has the World's Deepest Voice." *The Telegraph*, www.youtube.com/watch?v=8jCPl7Rcmmo. Accessed September 13, 2023.

Sissom, Dawn E. Frazer, D.A. Rice, and G. Peters. 1991. "How Cats Purr." *Journal of Zoology* 223(1): 67–78.

Smith, Ian. 2022. "Ukraine War: Europeans Rush to Buy Iodine Pills amid Fears of Nuclear Catastrophe" *Euronews*, March 7, 2022. www.euronews.com/health/2022/03/07/ukraine-war-european-pharmacies-face-jump-in-demand-for-iodine-pills-after-putin-s-nuclear. Accessed May 11, 2023.

Snively, E., A.P. Russell, G.L. Powell, J.M. Theodor, and M.J. Ryan. 2014. "The Role of the Neck in the Feeding Behaviour of the Tyrannosauridae: Inference Based on Kinematics and Muscle Function of Extant Avians." *Journal of Zoology* 292(4): 290–303.

Starnberger, Iris, Doris Preininger, and Walter Hödl. 2014. "The Anuran Vocal Sac: A Tool for Multimodal Signalling." *Animal Behaviour* 97: 281–288.

Stein, Judith. 1994. "Pippin." *Pennsylvania Heritage* (Spring). http://paheritage.wpengine.com/article/pippin/. Accessed February 12, 2022.

Steiner, Emily. 2022. "Neck Verse." *New Literary History* 53(3): 333–362.

Sterling, Michele, and Justin Kenardy. 2011.*Whiplash: Evidence Base for Clinical Practice.* Melbourne: Elsevier Australia.

Sterpetti, Antonio, Giorgio De Toma, and Alessandro De Cesare. 2015. "Thyroid Swellings in the Art of the Italian Renaissance." *American Journal of Surgery* 210(3): 591–596.

Stewart, Jude. 2013. "Why Are Jeans Blue?" *Slate*, October 14, 2013. https://slate.com/human-interest/2013/10/blue-jeans-what-s-the-reason-behind-the-color.html. Accessed August 22, 2023.

Sustaita, Diego, Margaret Rubega, and Susan Farabaugh. 2018. "Come on Baby, Let's Do the Twist: The Kinematics of Killing in Loggerhead Shrikes." *Biology Letters* 14(9): 20180321 (online).

Suthers, Roderick, Franz Goller, and Rebecca Hartley. 1994. "Motor Dynamics of Song Production by Mimic Thrushes." *Journal of Neurobiology* 25(8): 917–936.

Suthers, Roderick, Eric Vallet, and Michel Kreutzer. 2012. "Bilateral Coordination and the Motor Basis of Female Preference for Sexual Signals in Canary Song." *Journal of Experimental Biology* 215(17): 2950–2959.

Tauber, Yanki. N.d. "The Kabbalah of the Neck." *Chabad.* www.chabad.org/library/article_cdo/aid/363823/jewish/The-Kabbalah-of-the-Neck.htm. Accessed June 14, 2023.

Taylor, Michael P., and Mathew J. Wedel. 2013. "Why Sauropods Had Long Necks; and Why Giraffes Have Short Necks." *PeerJ* 1: e36 (online).

Taylor, Ryan, Barrett Klein, Joey Stein, and Michael Ryan. 2008. "Faux Frogs: Multimodal Signalling and the Value of Robotics in Animal Behaviour." *Animal Behaviour* 76(3): 1089–1097.

———. 2011. "Multimodal Signal Variation in Space and Time: How Important Is Matching a Signal with Its Signaler?" *Journal of Experimental Biology* 214(5): 815–820.

Teitlell, Beth. 2017. "Trump Wears His Ties Long. It's Not by Accident." *Boston Globe*, February 10, 2017.

Telford, Taylor. 2021. "Dozens of Nooses Have Shown Up on U.S. Construction Sites. The Culprits Rarely Face Consequences." *Washington Post*, July 22, 2021.

Theurer, Jessica. 2014. "Trapped in Their Own Rings: Padaung Women and Their Fight for Traditional Freedom." *International Journal of Gender and Women's Studies* 2(4): 51–67.

"Throat Singing." N.d. "Throat Singing: A Unique Vocalization from Three Cultures." Smithsonian Folkways Recordings, https://folkways.si.edu

/throat-singing-unique-vocalization-three-cultures/world/music/article/smithsonian. Accessed April 10, 2023.

Titze, Ingo. 2012. "Why Lions Roar like Babies Cry." *Physics World* 25(11): 52–53.

Titze, Ingo, and Anil Palaparthi. 2018. "Radiation Efficiency for Long-Range Vocal Communication in Mammals and Birds." *Journal of the Acoustical Society of America* 143(5): 2813–2824.

Tong, Darryl, and Ross Beirne. 2013. "Combat Body Armor and Injuries to the Head, Face, and Neck Region: A Systematic Review." *Military Medicine* 178(4): 421–426.

Toomer, Jean, Charles Johnson, and Rudolph P. Byrd. 1931. *Essentials*. Athens: University of Georgia Press.

Torre, Carlos, Alberto Ramos, Salim Dib, Alexandre Abreu, and Alejandro Chediak. 2019. "Anatomy of Obstructive Sleep Apnea: An Evolutionary and Developmental Perspective." *International Journal of Head and Neck Surgery* 10(4): 98–101.

T'sjoen, Guy, Griet De Cuypere, Stan Monstrey, Piet Hoebeke, F. Kenneth Freedman, Mahesh Appari, Paul-Martin Holterhus, John Van Borsel, and Martine Cools. 2011. "Male Gender Identity in Complete Androgen Insensitivity Syndrome." *Archives of Sexual Behavior* 40: 635–638.

Turnbull, Oliver, Victoria Lovett, Jackie Chaldecott, and Marilyn Lucas. 2014. "Reports of Intimate Touch: Erogenous Zones and Somatosensory Cortical Organization." *Cortex* 53: 146–154.

Twain, Mark. 1880. *A Tramp Abroad*. London: Chatto.

Tyer, Charles L. 1992. "Yoke." In *The Anchor Bible Dictionary*, edited by D.N. Freedman, 1026–1027. New Haven, CT: Yale University Press.

Uematsu, Sumio, Andrew Yang, Thomas Preziosi, Richard Kouba, and T.J. Toung. 1983. "Measurement of Carotid Blood Flow in Man and Its Clinical Application." *Stroke* 14: 256–266.

VanBuren, Collin, and David Evans. 2017. "Evolution and Function of Anterior Cervical Vertebral Fusion in Tetrapods." *Biological Reviews* 92(1): 608–626.

Vesalius, Andreas. 1998. *On the Fabric of the Human Body*. Translated by W.F. Richardson and J.B. Carman. Novato, CA: Norman Publishing.

Vincent, Susan J. 2009. *The Anatomy of Fashion: Dressing the Body from the Renaissance to Today*. Oxford: Berg Publishers.

Vogel, Steven. 2003. *Prime Mover: A Natural History of Muscle*. New York: W.W. Norton & Company.

Walker, T.J. 2011. "Move Your Head—Media Training." *Forbes*, February 28, 2011.

Wang, Shi-Qi, Jie Ye, Jin Meng, Chunxiao Li, Loïc Costeur, Bastien Mennecart, Chi Zhang, et al. 2022. "Sexual Selection Promotes Giraffoid Head-Neck Evolution and Ecological Adaptation." *Science* 376(6597): eabl8316 (online).

Watkinson, William. 2016. "UK Counter Terror Experts to Issue Guidelines Warning Priests of Possible Normandy-Style Attack." *International Business Times*, August 30, 2016. www.ibtimes.co.uk/uk-counter-terror-experts-issue-guidelines-warning-priests-possible-normandy-style-attack-1578813. Accessed July 11, 2022.

Wedel, Mathew J. 2011. "A Monument of Inefficiency: The Presumed Course of the Recurrent Laryngeal Nerve in Sauropod Dinosaurs." *Acta Palaeontologica Polonica* 57(2): 251–256.

West, Peyton. 2005. "The Lion's Mane." *American Scientist* 93(3): 226–235.

West, Peyton, and Craig Packer. 2002. "Sexual Selection, Temperature, and the Lion's Mane." *Science* 297(5585): 1339–1343.

White, Edward. 2018. "The Bloody Family History of the Guillotine." *Paris Review*, April 6, 2018. www.theparisreview.org/blog/2018/04/06/the-bloody-family-history-of-the-guillotine/. Accessed June 10, 2023.

Wickler, Wolfgang, and Uta Seibt. 1995. "Syntax and Semantics in a Zulu Bead Colour Communication System." *Anthropos* 90: 391–405.

Wickman, Forest. 2012. "Working Man's Blues." *Slate*, May 1, 2012. https://slate.com/business/2012/05/blue-collar-white-collar-why-do-we-use-these-terms.html. Accessed March 28, 2022.

Wilkinson, David M., and Graeme D. Ruxton. 2012. "Understanding Selection for Long Necks in Different Taxa." *Biological Reviews* 87(3): 616–630.

Williams, Tennessee. 1986. *Cat on a Hot Tin Roof: A Play in Three Acts*. Sewanee, TN: Dramatists Play Service.

Williquette, Heather Ann. 2019. "Investigating the Reasons Women Wear Choker Necklaces." PhD dissertation, University of Colorado.

Willmann, Magali, and Benoît Bolmont. 2012. "The Trapezius Muscle Uniquely Lacks Adaptive Process in Response to a Repeated Moderate Cognitive Stressor." *Neuroscience Letters* 506(1): 166–169.

Wolfe, Tom. 2002. *The Bonfire of the Vanities: A Novel*. New York: Farrar, Straus, and Giroux.

World Health Organization (WHO). 2005. "Chernobyl: The True Scale of the Accident. 20 Years Later a UN Report Provides Definitive Answers and Ways to Repair Lives." News release. www.who.int/news/item/05-09-2005-chernobyl-the-true-scale-of-the-accident. Accessed January 10, 2022.

Wynne-Jones, Jonathan. 2007. "Vicars Urged to Drop 'Risky' Dog Collars." *The Telegraph*, October 7, 2007.

Xydakis, Michael, Michael Fravell, Katherine Nasser, and John Casler. 2005. "Analysis of Battlefield Head and Neck Injuries in Iraq and Afghanistan." *Otolaryngology—Head and Neck Surgery* 133(4): 497–504.

Yeivin, Ze'ev, and Louis Isaac Rabinowitz. 2007. "Yoke." In *Encyclopaedia Judaica*, 2nd ed., vol. 21, edited by M. Berenbaum and F. Skolnik, 381. Detroit, MI: Macmillan Reference USA.

Young, Bruce, and Kenneth Kardong. 2010. "The Functional Morphology of Hooding in Cobras." *Journal of Experimental Biology* 213(9): 1521–1528.

Young, Robb. 2013. "Christine Lagarde: Dressing All the Way to the Bank." *BBC*, May 1, 2013.

Zamponi, Virginia, Rossella Mazzilli, Fernando Mazzilli, and Marco Fantini. 2021. "Effect of Sex Hormones on Human Voice Physiology: From Childhood to Senescence." *Hormones* 20(4): 691–696.

Zamudio, Kelly, and Barry Sinervo. 2000. "Polygyny, Mate-Guarding, and Posthumous Fertilization as Alternative Male Mating Strategies." *Proceedings of the National Academy of Sciences* 97(26): 14427–14432.

Zheng, Liying, Gunter Siegmund, Gulsum Ozyigit, and Anita Vasavada. 2013. "Sex-Specific Prediction of Neck Muscle Volumes." *Journal of Biomechanics* 46(5): 899–904.

Zilczer, Judith. 2001. "A Not-So-Peaceable Kingdom: Horace Pippin's 'Holy Mountain.'" *Archives of American Art Journal* 41: 18–33.

Zimmermann, Michael B. 2008. "Research on Iodine Deficiency and Goiter in the 19th and Early 20th Centuries." *Journal of Nutrition* 138: 2060–2063.

Zimmermann, Michael B., and Maria Andersson. 2021. "Global Perspectives in Endocrinology: Coverage of Iodized Salt Programs and Iodine Status in 2020." *European Journal of Endocrinology* 185(1): R13–R21.

Zimmermann, Michael B., Pieter L. Jooste, and Chandrakant S. Pandav. 2008. "Iodine-Deficiency Disorders." *The Lancet* 372: 1251–1262.

INDEX

Founded in 1893,
UNIVERSITY OF CALIFORNIA PRESS
publishes bold, progressive books and journals on topics in the arts, humanities, social sciences, and natural sciences—with a focus on social justice issues—that inspire thought and action among readers worldwide.

The UC PRESS FOUNDATION
raises funds to uphold the press's vital role as an independent, nonprofit publisher, and receives philanthropic support from a wide range of individuals and institutions—and from committed readers like you. To learn more, visit ucpress.edu/supportus.